# Essential Statistical Physics

This clear and pedagogical text delivers a concise overview of classical and quantum statistical physics. *Essential Statistical Physics* shows students how to relate the macroscopic properties of physical systems to their microscopic degrees of freedom, preparing them for graduate courses in areas such as biophysics, condensed matter physics, atomic physics and statistical mechanics. Topics covered include the microcanonical, canonical, and grand canonical ensembles, Liouville's theorem, kinetic theory, non-interacting Fermi and Bose systems and phase transitions, and the Ising model. Detailed steps are given in mathematical derivations, allowing students to quickly develop a deep understanding of statistical techniques. End-of-chapter problems reinforce key concepts and introduce more advanced applications, and appendices provide a detailed review of thermodynamics and related mathematical results. This succinct book offers a fresh and intuitive approach to one of the most challenging topics in the core physics curriculum, and provides students with a solid foundation for tackling advanced topics in statistical mechanics.

**Malcolm P. Kennett** is an associate professor at Simon Fraser University, British Columbia. He studied at the University of Sydney and Princeton University and was a postdoctoral fellow at the University of Cambridge. He has taught statistical mechanics at both undergraduate and graduate level for many years and has been recognized for the high quality of his teaching and innovative approaches to the undergraduate and graduate curriculum. His research is focused on condensed matter theory and he has made contributions to the theory of spin glasses, dilute magnetic semiconductors, out-of-equilibrium dynamics in ultracold atoms, and the quantum Hall effect in graphene.

# Essential Statistical Physics

MALCOLM P. KENNETT

Simon Fraser University, British Columbia

CAMBRIDGE
UNIVERSITY PRESS

# CAMBRIDGE
## UNIVERSITY PRESS

University Printing House, Cambridge CB2 8BS, United Kingdom

One Liberty Plaza, 20th Floor, New York, NY 10006, USA

477 Williamstown Road, Port Melbourne, VIC 3207, Australia

314–321, 3rd Floor, Plot 3, Splendor Forum, Jasola District Centre, New Delhi – 110025, India

79 Anson Road, #06–04/06, Singapore 079906

Cambridge University Press is part of the University of Cambridge.

It furthers the University's mission by disseminating knowledge in the pursuit of education, learning, and research at the highest international levels of excellence.

www.cambridge.org
Information on this title: www.cambridge.org/9781108480789
DOI: 10.1017/9781108691116

First published 2021

Printed in the United Kingdom by TJ International Ltd, Padstow Cornwall, 2021

*A catalogue record for this publication is available from the British Library.*

ISBN 978-1-108-48078-9 Hardback

# Contents

# Preface

The ideas of statistical physics allow one to relate macroscopic properties of systems to their microscopic degrees of freedom and play important roles in many branches of physics and other sciences, and disciplines further afield such as economics. In particular, the concept of entropy as a measure of ignorance has found very wide application, especially recently in its relation to information theory. Statistical physics is hence one of the core courses in many undergraduate degree programs, sometimes taught as a combined course with thermal physics, but often as a course in its own right.

As an instructor for an upper-level undergraduate course in statistical physics, I have found it challenging to find a text for the course that contains the important ideas of statistical physics at a level the students can relate to, and at an affordable price. My goal in writing this book has been to cover key concepts and examples in statistical physics in a pedagogical and clear manner, to include mathematical detail in a way that is useful to upper-level undergraduates and beginning graduate students, while also being concise and affordable. This necessarily leads to a limited scope. Unlike many existing texts on statistical physics, I do not try to cover thermal physics and statistical mechanics in the same book, nor do I try to be comprehensive in covering both undergraduate and graduate statistical mechanics in the same volume. My goal is that this book is one that will provide students and instructors with the essential ideas of statistical physics in a way that will prove useful to all students. For students continuing to graduate studies in physics and related topics I aim to give a comprehensive coverage of ideas that are needed in graduate school, and for students who do not continue to graduate studies I aim to give a solid background in statistical physics that will allow them to apply these ideas in other contexts.

The book is based on lecture notes for a one-semester (13 weeks, 3 hours of lectures/week) undergraduate course on statistical physics that I have delivered three times at Simon Fraser University (SFU). I have assumed that the reader of the book has already taken (or is taking at the same time) a course in thermal physics, so I do not elaborate greatly on thermodynamic quantities, although I have included a brief primer on thermal physics in Appendix B, so that the reader can look up ideas as needed. I also assume that the reader has taken at least one quantum mechanics course, so they are familiar with the solution of quantum mechanical problems like the particle in a box and the simple harmonic oscillator. My target audience for this book is primarily undergraduate students learning statistical mechanics for the first time. However, I also believe that it will be of interest to graduate students wanting a clear reference for statistical mechanics that provides details of calculations and clarification of concepts, and to instructors looking for an affordable, well-structured text or reference for statistical physics courses.

My choice of content is what I consider to be the main ideas and examples of statistical physics that will allow a starting graduate student in physics to have a solid foundation in the subject, but is not intended to be exhaustive. Hence, after an introduction to probability, microstates and macrostates (Chapter 1), I cover the microcanonical (Chapter 2), canonical (Chapter 4) and grand canonical (Chapter 6) ensembles. I take the opportunity to discuss Liouville's theorem and ergodicity after discussing the microcanonical ensemble (Chapter 3) and give an introduction to kinetic theory after the canonical ensemble (Chapter 5). After discussing the grand canonical ensemble, I introduce quantum statistical mechanics (Chapter 7) and devote a chapter each to fermions (Chapter 8) and bosons (Chapter 9), touching on Bose–Einstein condensation and superfluidity and applications of Fermi statistics to metals and compact stars. Finally, I finish with a brief introduction to phase transitions, broken symmetry and ordering, using the Ising model as the main example, and introduce Landau's theory of phase transitions (Chapter 10). There are problems at the end of all the chapters to help reinforce the concepts discussed in the bodies of the chapters.

I had had a vague notion that at some point I would like to write the book here for a number of years; almost certainly nothing would have come of these ideas but for a meeting with Nicholas Gibbons of Cambridge University Press in 2015. Nicholas encouraged me to flesh out the lecture notes that I had prepared for the course into a text. It has taken several years for that to happen, but I hope you find the end result useful.

In addition to Nicholas Gibbons, I would like to thank everyone at Cambridge University Press who I have interacted with on this project, particularly Liso Pinto, Maggie Jeffers and Rachel Norridge who have helped keep me on track and provided much helpful advice. I would also like to thank Jeff McGuirk and Michael Plischke from SFU, who have shared their experience of teaching statistical mechanics with me, the students at SFU who have provided me with considerable feedback on early drafts of this book, specifically Florian Baer, Matt Wiens, Aidan Wright, Frank Wu and Adrian Yeung, and my father Brian Kennett for detailed feedback on all aspects of the book. Finally, I would like to thank my wife Kaila and my children Heath and Eily for their love, support and patience during the writing of this book, especially in the later stages of writing.

# 1 Introduction

## 1.1 What is Statistical Mechanics?

We perceive the world at a macroscopic level, dealing with substances composed of atoms, yet wholly ignorant of the state of those individual atoms. As physicists, there are many situations in which we may wish to describe the macroscopic behaviour of some physical system, even though we do not have full knowledge of the state of all of the microscopic degrees of freedom of that system. For instance, suppose we want to know the physical properties of a container of some gas, whose atoms barely interact with each other, such as argon. The most brute-force approach to trying to understand the physical properties of the gas would be to take the Schrödinger equation for the $\sim 10^{23}$ atoms in the container, and try to solve it to find the eigenvalues and eigenstates of any observable that we might be interested in. Even if we were able to do this (which we aren't), we would still need to know the initial state of the gas in order to be able to determine the state of the gas at any particular time. Moreover, the subsequent dynamics will be chaotic, so that we can never know the initial conditions with sufficient precision to be able to calculate the state of the gas at arbitrary times. Clearly, such an approach is impractical, even with the fastest computers currently available (or likely to be available in the future).

One of the triumphs of eighteenth- and nineteenth-century physics was the development of thermodynamics, which allowed the description of the physical properties of a system such as a container of gas to be reduced to a single equation of state, the ideal gas law

$$PV = Nk_BT, \tag{1.1}$$

which relates thermodynamic quantities pressure $P$, volume $V$ and temperature $T$ without any reference to the microscopic details of the gas.

However, thermodynamics does not explain why, even though the ideal gas law does a good job of describing dilute gases of both argon and oxygen (or any other simple gas), argon has a heat capacity $C_V = \frac{3}{2}k_B$, whilst oxygen has a heat capacity $C_V = \frac{5}{2}k_B$. The distinction between these two heat capacities depends on the microscopic degrees of freedom that are available to atoms or molecules in each gas. As diatomic molecules, oxygen molecules have extra degrees of freedom than argon atoms, which leads to more possible ways for energy to be deposited in such a gas. For instance, oxygen molecules can have rotational kinetic energy in addition to the translational kinetic energy that argon atoms can have. Statistical mechanics is a framework to calculate such macroscopic properties, from a knowledge of microscopic details of a system. Researchers such as Maxwell, Boltzmann and Gibbs developed these ideas in a classical context, but they have proven to be equally applicable to quantum systems.

While we study a number of problems in this book that can be solved exactly, most problems in statistical physics cannot be solved explicitly. Nevertheless, the concepts that we develop here can be applied either in numerical contexts, such as molecular dynamics or Monte Carlo simulations, or in approximate analytic approaches. Statistical mechanics also allows us to understand why we can describe the properties of a complex system with many different possible arrangements of its microscopic constituents, for example $\sim 10^{23}$ atoms, in terms of just a few quantities, such as temperature, pressure and chemical potential, and why thermal equilibrium exists. As such, it provides a microscopic foundation to the thermodynamic concepts used in physics and other fields such as chemistry, engineering and biology.

An important part of thermodynamics is the concept of entropy, which is often described in somewhat unsatisfying terms as "the amount of disorder in a system" or some similar phrase. In the context of statistical mechanics, entropy can be given a precise meaning, as a measure of our ignorance about a system. Such a view is in accord with the idea that we describe the statistical properties of a system rather than trying to solve the equations of motion for its constituents (e.g. Newton's laws or the Schrödinger equation). More broadly, the idea of entropy also plays an important role in information theory. In Chapter 2 we will see how entropy is related to irreversibility, as embodied in the second law of thermodynamics, even though the relevant equations of motion are invariant under time reversal.

Examples of systems and situations that we will encounter and/or can be described in a statistical mechanical framework include: ideal gases, spins in magnetic materials, photons in the Sun, white dwarf stars, molecular motors, DNA, Bose–Einstein condensation, impurities in crystals, polymer folding, stock market fluctuations, electrical noise in circuits, magnetic domains, superfluidity and phase transitions.

## 1.2 Probabilistic Behaviour

Having established that it is futile to try to follow the trajectories of every single molecule in a gas or to solve the corresponding equations of motion for the particles, we turn to a statistical description that will allow us to obtain the collective behaviour of the system without knowing what any individual particle is doing. This is in accord with how we describe physical systems – we generally are interested in macroscopic properties, such as pressure or temperature in a gas, that arise from the average over the motion of individual molecules. We know that certain configurations of the molecules are very unlikely, e.g. all the molecules being found in one corner of a container. To quantify this intuitive understanding we will assign probabilities to different configurations. Averages over these probabilities will give us the average values of physical quantities, which we will be able to relate to likely outcomes of macroscopic measurements when the number of constituents of a system becomes large. To give such a description, we need to start with a careful discussion of probabilities and probability distributions.

There are a number of different ways to think about probability. An empirical approach is the frequentist point of view: if we do $N$ trials of an experiment and measure $X$ for some quantity in $N_X$ of those trials, then the frequentist view is that the probability of measuring $X$ is

$$P(X) = \lim_{N \to \infty} \frac{N_X}{N}. \tag{1.2}$$

A problem with this approach is that $N \to \infty$ is an unattainable limit, it is impossible to make an infinite number of measurements – it also ties a probability to an actual sequence of events. Another way to think about probability is via a propensity approach, in which the probability is viewed as intrinsic to a particular physical situation and is then reflected in subsequent measurements. In this view, $P(X)$ exists for the physical situation we are considering and then if we make a sequence of measurements, those measurements will be determined by $P(X)$. A third approach is the Bayesian point of view, where the probabilities assigned to different possible outcomes, e.g. $X$ or $Y$, depend on prior knowledge. We will tend to take a propensity approach to probabilities, but from a mathematical point of view, provided a probability satisfies the axioms, then it is acceptable.

## 1.2.1 Axioms of Probability

We introduce $P(A)$ as the probability that some outcome $A$ occurs given some initial conditions, e.g. this could be the probability that a fair coin comes up heads when tossed. In order to be a probability, $P(A)$ must satisfy the following four axioms, which express basic properties that we require of any reasonable probability:

**(1)** $P(A) \geq 0$ – this guarantees there are no negative probabilities.
**(2)** $\sum_i P(A_i) = 1$ – this expresses the idea that if we consider all possible outcomes $A_i$ for the given initial conditions then one of these must happen and so the sum of all the probabilities is 1.
**(3)** The probability when combining independent outcomes for the same event is additive (e.g. this could be the probability of rolling either a 1 or a 4 on a die, in which case the probability is equal to the sum of the probability of rolling 1 with the probability of rolling 4). Hence for independent outcomes $A$ and $B$

$$P(A \lor B) = P(A) + P(B), \tag{1.3}$$

where $\lor$ is logical *or*, or equivalently

$$P(A) = 1 - P(\bar{A}), \tag{1.4}$$

where $\bar{A}$ is logical *not A*. This is also equivalent to

$$P(A) + P(\bar{A}) = 1, \tag{1.5}$$

which reflects that it is certain that we get one of the outcomes $A$ or $\bar{A}$.
**(4)** The probability of independent events is multiplicative (e.g. we might want to know the probability that we get a head when we flip a coin and a 6 when we roll a die):

$$P(A \land B) = P(A)P(B), \tag{1.6}$$

where $\land$ is logical *and*.

## 1.2.2 Example: Coin Toss Experiment

A simple example which illustrates the ideas expressed in the axioms in a concrete way, and which also introduces the binomial distribution, is a coin toss experiment. We can flip a coin many times and we will get some sequence of heads and tails, e.g.

HTTHHHTHTTTTTHTHTHTHTHTHHTHTTHTHTHTHTHTHTHTTTHHTH...

We can't predict the outcome of any one coin toss, but for a fair coin, the odds of having a head or a tail when we toss the coin are $p = \frac{1}{2}$ and $q = \frac{1}{2}$, respectively.

For $N$ tosses of the coin, we can determine the probability of there being $N_H$ heads and $N_T$ tails by drawing a tree diagram as illustrated in Fig. 1.1 and counting possibilities, for instance, or more systematically by using the binomial theorem:

$$1 = (p + q)^N = p^N + Np^{N-1}q + \frac{1}{2}N(N-1)p^{N-2}q^2 + \cdots + q^N$$

$$= \sum_{M=0}^{N} \binom{N}{M} p^{N-M}q^M$$

$$= \sum_{M=0}^{N} P(N, N - M), \tag{1.7}$$

where

$$\binom{N}{M} = \frac{N!}{M!\,(N-M)!}, \tag{1.8}$$

and $P(N, N - M)$ is the probability of having $N - M$ heads and $M$ tails. If we label the number of heads as $N_H$ and the number of tails as $N_T$, then the probability of having $N_H$ heads and $N_T$ tails after a total of $N = N_H + N_T$ coin tosses will be given by

$$P(N, N - M) = P(N, N_H) = \binom{N}{M} p^{N-M}q^M = \frac{N!}{N_H!\,N_T!}p^{N_H}q^{N_T}. \tag{1.9}$$

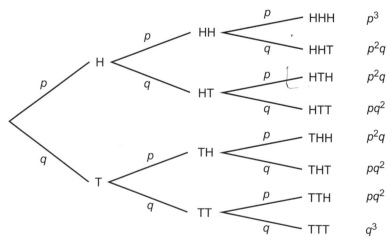

**Fig. 1.1**　Probability tree diagram for three coin tosses. The probability of obtaining a head (H) on any coin toss is $p$ and the probability of obtaining a tail (T) is $q$.

Now, suppose we pick $N = 100$, then our general expectation based on experience is that $\Delta N = N_H - N_T$ should not be too large (i.e. much less than 100). We can be quantitative and we see that the probability that we get $\Delta N = \pm 100$ is $p^{100} = 1/2^{100} \simeq 8 \times 10^{-31}$. In contrast, we might guess that it is quite likely that we will get $\Delta N = 0$, which is given by the probability of 50 heads:

$$P(100, 50) = \binom{100}{50} \frac{1}{2^{100}} \simeq 8.0\%. \tag{1.10}$$

However, this is not very different from the probability of $\Delta N = \pm 2$, corresponding to 49 or 51 heads:

$$P(100, 51) = P(100, 49) = \binom{100}{49} \frac{1}{2^{100}} \simeq 7.8\%, \tag{1.11}$$

but considerably more than $\Delta N = \pm 20$, which corresponds to 40 or 60 heads:

$$P(100, 40) = P(100, 60) = \binom{100}{40} \frac{1}{2^{100}} \simeq 1.1\%. \tag{1.12}$$

If we were to perform $N = 10$ coin flips, for instance, then looking at $\Delta N$ as we have done here would not be very useful to compare with what we find when $N = 100$. A quantity that is more useful is

$$\frac{\Delta N}{N} = \frac{N_H - N_T}{N},$$

which gives us the relative deviation from the behaviour expected of a fair coin (i.e. $N_H = N_T$). We might expect that the probability that this quantity is close to zero gets larger as $N$ increases. For instance, when $N = 100$, the probability that $-0.1 \leq \Delta N/N < 0.1$ is $\simeq 73\%$, whereas for $N = 1000$, the probability is $\simeq 99.9\%$.

### 1.2.3 Probability Distributions

The coin toss example suggests that rather than focusing on individual outcomes of events, which are random, we should look at quantities that are robust properties of the system, such as average values or the distribution of allowed values – we may not know the exact result of any particular event, but would like to know the range of highly probable outcomes.

The objects that will be central to this discussion are probability distributions. A probability distribution $P(X)$ for some quantity $X$ gives the probability of the different allowed values of $X$. It can be either discrete, in which case it will take the form $p(x_i)$, which measures the probability that $X$ takes the value $x_i$, or continuous, in which case we can write the probability as $p(x)dx$, which corresponds to the probability that $X$ takes a value between $x$ and $x + dx$. In this second case, $p(x)$ is a probability density, as illustrated in Fig. 1.2.

Given a probability distribution $p(x_i)$, we can calculate the mean value of some function $f$ of $X$ by weighting the values of $f$ by the probability distribution:

$$\langle f(x) \rangle = \overline{f(x)} = \sum_i f(x_i)\, p(x_i). \tag{1.13}$$

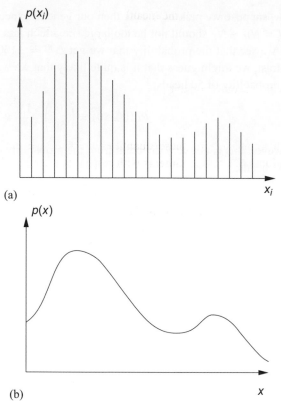

(a)

(b)

**Fig. 1.2**    Examples of (a) a discrete and (b) a continuous probability distribution.

For a continuous probability distribution we should replace the sum with an integral:

$$\langle f(x) \rangle = \int dx \, f(x) \, p(x). \qquad (1.14)$$

In the following, results written with summations should be understood as applying to both discrete and continuous probability distributions. We will be particularly interested in moments of probability distributions as these will allow us to characterize the shape of unknown probability distributions. For instance, for the $m^{\text{th}}$ moment:

$$\langle x^m \rangle = \overline{x^m} = \sum_i x_i^m \, p(x_i). \qquad (1.15)$$

The main moments of distributions that we will be concerned with are:

(i) the zeroth moment

$$\left\langle x^0 \right\rangle = \sum_i p(x_i) = 1 \qquad (1.16)$$

(as we would expect for a probability);

(ii) the first moment (the mean)

$$\langle x \rangle = \bar{x} = \sum_i x_i \, p(x_i), \tag{1.17}$$

and;

(iii) the second moment

$$\langle x^2 \rangle = \sum_i x_i^2 \, p(x_i). \tag{1.18}$$

With knowledge of these quantities we can also determine the central moments of a distribution, i.e. the moments with respect to the mean. Define

$$\Delta x = x - \langle x \rangle,$$

then we can see immediately that

$$
\begin{aligned}
\langle \Delta x \rangle &= \langle x - \langle x \rangle \rangle \\
&= \sum_i x_i \, p(x_i) - \sum_i \langle x \rangle \, p(x_i) \\
&= \langle x \rangle - \langle x \rangle \, 1 \\
&= 0,
\end{aligned}
\tag{1.19}
$$

and that the variance is

$$
\begin{aligned}
\langle (\Delta x)^2 \rangle &= \langle (x - \langle x \rangle)^2 \rangle \\
&= \langle x^2 \rangle - 2 \langle x \langle x \rangle \rangle + \langle x \rangle^2 \\
&= \langle x^2 \rangle - 2 \langle x \rangle^2 + \langle x \rangle^2 \\
&= \langle x^2 \rangle - \langle x \rangle^2,
\end{aligned}
\tag{1.20}
$$

which is closely related to the standard deviation

$$\sigma = \sqrt{\langle (\Delta x)^2 \rangle} = \sqrt{\langle x^2 \rangle - \langle x \rangle^2}. \tag{1.21}$$

We note in passing that the third and fourth central moments give the skew and the kurtosis, respectively, which can be used to characterize the shape of a probability distribution. These are important quantities in the theory of probability distributions but we will not use them in our discussion of statistical physics.

## 1.2.4 Example: Random Walk

Random walks describe many processes that can take place in nature. A very short list of examples includes the folding of polymers, Brownian motion of small particles, photons diffusing in the Sun, electrons diffusing in a wire, molecular motors, chemotaxis, genetic drift and stock prices. A schematic example of a random walk in two dimensions is illustrated in Fig. 1.3.

We will initially consider the simplest version of a random walk. Imagine a walker placed on a one-dimensional line. The walker takes steps of equal length randomly either to the left or right. Without knowing the exact sequence of steps that the walker takes, we

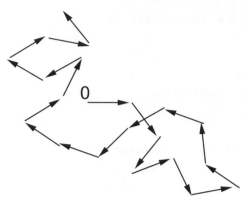

**Fig. 1.3**  A random walk in two dimensions with fixed step length but random direction for each step starting from the origin 0.

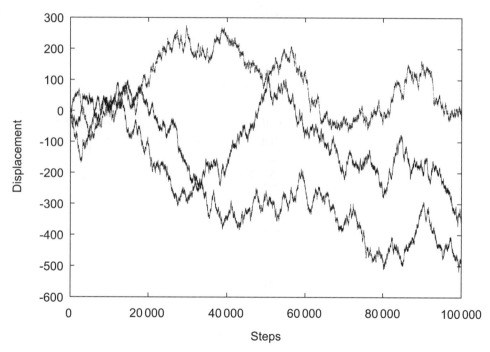

**Fig. 1.4**  Three random walks in one dimension, each with 100 000 steps.

cannot determine exactly where they will end up (see Fig. 1.4). However, we can say very precisely what the probabilities for finding the walker at any position along the line are after the walker has taken $N$ steps. Note that we could regard our coin-flipping example as a random walk, with heads corresponding to a move to the right and tails to a move to the left, so we are also obtaining the probability distribution for coin flips.

To simplify the problem, suppose the walker starts at the origin, and takes steps of length $a$. After they have taken $N$ steps, $N_L$ will have been to the left and $N_R$ will have been to the right, and let us call their position on the line $Ra$. Then

$$N = N_R + N_L, \tag{1.22}$$

$$R = N_R - N_L, \tag{1.23}$$

and so

$$N_R = \frac{1}{2}(N + R), \tag{1.24}$$

$$N_L = \frac{1}{2}(N - R). \tag{1.25}$$

There are two possibilities at each step, a move to the right with probability $p$ and a move to the left with probability $q$. In this sense the problem is just like the coin flip problem, with $N_R$ equivalent to $N_H$ and $N_L$ equivalent to $N_T$. Hence, for a walk with displacement from the origin of $Ra$, we can determine from the binomial theorem that the probability of any path with $N$ steps of which $N_R$ are to the right and $N_L$ are to the left is

$$P(N, R) = \frac{N!}{N_R! \, N_L!} p^{N_R} q^{N_L} = \frac{N!}{\left[\frac{1}{2}(N+R)\right]! \left[\frac{1}{2}(N-R)\right]!} p^{\left[\frac{1}{2}(N+R)\right]} q^{\left[\frac{1}{2}(N-R)\right]}. \tag{1.26}$$

The probability of each individual path is $p^{N_R} q^{N_L}$ (where the combinatorial factor $N!/N_R! \, N_L!$ takes care of the fact that there are multiple paths that end up at the same endpoint).

If we want to calculate the mean displacement for a walk with $N$ steps, then we can do so as follows (recalling that $N_R - N_L = N_R - [N - N_R]$):

$$\begin{aligned}
\langle R \rangle &= \sum_{R=-N}^{N} R \, P(N, R) \\
&= \sum_{N_R=0}^{N} (N_R - N_L) \frac{N!}{N_R! \, (N - N_R)!} p^{N_R} q^{(N - N_R)} \\
&= p \frac{\partial}{\partial p} \left[ \sum_{N_R=0}^{N} \frac{N!}{N_R! \, (N - N_R)!} p^{N_R} q^{(N - N_R)} \right]\Bigg|_{p+q=1} \\
&\quad - q \frac{\partial}{\partial q} \left[ \sum_{N_R=0}^{N} \frac{N!}{N_R! \, (N - N_R)!} p^{N_R} q^{(N - N_R)} \right]\Bigg|_{p+q=1} \\
&= p \frac{\partial}{\partial p} (p + q)^N \Big|_{p+q=1} - q \frac{\partial}{\partial q} (p + q)^N \Big|_{p+q=1} \\
&= Np (p + q)^{N-1} \Big|_{p+q=1} - Nq (p + q)^{N-1} \Big|_{p+q=1} \\
&= N(p - q). \tag{1.27}
\end{aligned}$$

The trick we used in order to evaluate this sum was to write the sum in terms of derivatives of the binomial expansion

$$(p + q)^N = \sum_{M=0}^{N} \binom{N}{M} p^{N-M} q^M, \tag{1.28}$$

where we note that $p$ and $q$ are continuous variables, and wait until the end of the calculation to set $p + q = 1$. If $p = q$ then we can see that $\langle R \rangle = 0$. We should expect this on general grounds without doing any calculations, because there is no asymmetry between right and left. This is because for any path that leads to a positive displacement, there will be an equally probable path that leads to the same negative displacement and the average of these displacements will be zero.

An alternative physically motivated argument that reaches the same value for $\langle R \rangle$ is that if the probability of a step to the right is $p$, then after $N$ steps we should expect that the number of steps to the right will be $\langle N_R \rangle = pN$. Similiarly, we should expect that the number of steps to the left will be $\langle N_L \rangle = qN$. Given these two estimates we would expect that $\langle R \rangle = \langle N_R \rangle - \langle N_L \rangle = (p - q)N$, which is exactly what we found in a more complicated calculation in deriving Eq. (1.27).

We can use a similar approach to the one we used in Eq. (1.27) to calculate the mean square displacement:

$$
\begin{aligned}
\langle R^2 \rangle &= \sum_{R=-N}^{N} R^2 P(N, R) \\
&= \sum_{N_R=0}^{N} (N_R - N_L)^2 \frac{N!}{N_R!\,(N - N_R)!} p^{N_R} q^{(N - N_R)} \\
&= \left( p \frac{\partial}{\partial p} \right)^2 \left[ \sum_{N_R=0}^{N} \frac{N!}{N_R!\,(N - N_R)!} p^{N_R} q^{(N - N_R)} \right]\Bigg|_{p+q=1} \\
&\quad - 2pq \frac{\partial^2}{\partial p \partial q} \left[ \sum_{N_R=0}^{N} \frac{N!}{N_R!\,(N - N_R)!} p^{N_R} q^{(N - N_R)} \right]\Bigg|_{p+q=1} \\
&\quad + \left( q \frac{\partial}{\partial q} \right)^2 \left[ \sum_{N_R=0}^{N} \frac{N!}{N_R!\,(N - N_R)!} p^{N_R} q^{(N - N_R)} \right]\Bigg|_{p+q=1} \\
&= \left\{ \left( p \frac{\partial}{\partial p} \right)^2 (p + q)^N - 2pq \frac{\partial^2}{\partial p \partial q} (p + q)^N + \left( q \frac{\partial}{\partial q} \right)^2 (p + q)^N \right\}\Bigg|_{p+q=1} \\
&= \left\{ p \frac{\partial}{\partial p} pN(p + q)^{N-1} - 2pqN(N - 1)(p + q)^{N-2} + q \frac{\partial}{\partial q} qN(p + q)^{N-1} \right\}\Bigg|_{p+q=1} \\
&= \Big[ Np(p + q)^{N-1} + p^2 N(N - 1)(p + q)^{N-2} - 2pqN(N - 1)(p + q)^{N-2} \\
&\quad + Nq(p + q)^{N-1} + q^2 N(N - 1)(p + q)^{N-2} \Big]\Big|_{p+q=1} \\
&= (p - q)^2 N(N - 1) + N \\
&= N(p + q)^2 + (p - q)^2 N(N - 1) \\
&= N^2 (p - q)^2 + 4Npq \\
&= \langle R \rangle^2 + 4Npq.
\end{aligned}
\tag{1.29}
$$

Hence the variance in the distance travelled in the walk is

$$\sigma_R^2 = \left\langle R^2 \right\rangle - \left\langle R \right\rangle^2 = 4Npq. \tag{1.30}$$

We see that the two terms in the expression for $\left\langle R^2 \right\rangle$ scale differently with $N$, one growing as $N$ and the other as $N^2$. If we have an unbiased random walk with $p = q = \frac{1}{2}$ then, as we saw above, $\langle R \rangle = 0$, and we can use Eq. (1.29) to determine the root mean squared (rms) distance from the origin after $N$ steps as a characteristic length scale for the walk:

$$L_{\mathrm{rms}} = a\sqrt{\langle R^2 \rangle} = \sqrt{N}\, a. \tag{1.31}$$

While we derived this result in the context of a one-dimensional random walk, it holds more generally in higher dimensions. If we imagine taking steps equally spaced in time, so that $N \propto t$, then we find that the distance from the origin for a random walk scales as $L_{\mathrm{rms}} \propto \sqrt{t}$, which is the characteristic behaviour of a diffusion process – the continuum limit of a random walk. In contrast, if $p \neq q$, then as $N$ gets large, the $N^2(p-q)^2$ term will dominate $\left\langle R^2 \right\rangle$. If we again take $N \propto t$, then we see that the rms length scale is $L_{\mathrm{rms}} = a\sqrt{\langle R^2 \rangle} \propto N \propto t$, so the displacement grows linearly with time, which is characteristic of motion in a straight line and is known as ballistic motion. In this case, at large times the bias in the walk is much more important than diffusion in determining the position of the walker.

### 1.2.5 Large-$N$ Limit of the Binomial Distribution

In the binomial distribution that arises in the coin-flipping problem and the random walk, the probability

$$P(N, R) = \frac{N!}{\left[\frac{1}{2}(N+R)\right]!\left[\frac{1}{2}(N-R)\right]!} p^{\left[\frac{1}{2}(N+R)\right]} q^{\left[\frac{1}{2}(N-R)\right]} \tag{1.32}$$

contains several factorials. It is possible to calculate factorials exactly for $N$ not too large, but when $N \sim 100$ this is already too large for a pocket calculator, so it is useful to be able to write factorials in another way that allows us to obtain accurate numerical results. A suitable way is Stirling's formula,[1] which is valid for large $N$ (in practice it is quite accurate even for relatively small values of $N$, e.g. for $N = 3$ it gives an error of about 3%):

$$N! \simeq \sqrt{2\pi N}\, N^N e^{-N}, \tag{1.33}$$

or

$$\ln(N!) \simeq \left(N + \frac{1}{2}\right)\ln(N) - N + \frac{1}{2}\ln(2\pi). \tag{1.34}$$

For small values of $N$, we have seen that the binomial distribution is reasonably sharply peaked (the coin toss experiment is an example), so we will start by trying to find the maximum of the distribution. It is more convenient to do this by working with the logarithm

---

[1] The details of the derivation of this result are presented in Appendix A.

of the distribution (since ln is a monotonic function, a maximum in $P$ will correspond to a maximum in $\ln P$):

$$\ln P(N, R) = \ln N! - \ln \left[ \left( \frac{1}{2}(N + R) \right)! \right] - \ln \left[ \left( \frac{1}{2}(N - R) \right)! \right]$$

$$+ \frac{1}{2}(N + R) \ln p + \frac{1}{2}(N - R) \ln q. \qquad (1.35)$$

To find the maximum for a given $N$, we treat the probability distribution as a continuous function of $R$, since the relative change in $\ln \left[ \left( \frac{1}{2}(N \pm R) \right)! \right]$ when $N \rightarrow N + 1$ is small when $N$ is large. Hence we need

$$\frac{\partial}{\partial R} \ln P(N, R) = \frac{1}{P} \frac{\partial}{\partial R} P(N, R) = 0. \qquad (1.36)$$

Now,

$$\frac{\partial}{\partial R} \ln P(N, R) = \frac{1}{2} \ln p - \frac{1}{2} \ln q - \frac{\partial}{\partial R} \left\{ \ln \left[ \left( \frac{1}{2}(N + R) \right)! \right] + \ln \left[ \left( \frac{1}{2}(N - R) \right)! \right] \right\},$$

$$(1.37)$$

and we can evaluate the derivatives of the factorials by replacing the factorials with Stirling's formula (assuming $N$ is large) so that the derivative term becomes

$$\simeq \frac{\partial}{\partial R} \left\{ \left[ \frac{1}{2}(N + R + 1) \right] \ln \left[ \frac{1}{2}(N + R) \right] - \left[ \frac{1}{2}(N + R) \right] + \frac{1}{2} \ln(2\pi) \right.$$

$$+ \left[ \frac{1}{2}(N - R + 1) \right] \ln \left[ \frac{1}{2}(N - R) \right] - \left[ \frac{1}{2}(N - R) \right] + \frac{1}{2} \ln(2\pi) \right\}$$

$$= \frac{1}{2} \ln \left[ \frac{1}{2}(N + R) \right] - \frac{1}{2} + \frac{1}{2} \frac{(N + R + 1)}{(N + R)} - \frac{1}{2} \ln \left[ \frac{1}{2}(N - R) \right] + \frac{1}{2} - \frac{1}{2} \frac{(N - R + 1)}{(N - R)}$$

$$= \frac{1}{2} \ln \left( \frac{N + R}{N - R} \right) + \frac{1}{2} \left[ \frac{1}{N + R} - \frac{1}{N - R} \right],$$

and hence in the large-$N$ limit where the terms $1/(N \pm R)$ are small compared to the logarithm term

$$\frac{\partial}{\partial R} \ln P(N, R) \simeq \frac{1}{2} \ln \frac{p}{q} - \frac{1}{2} \ln \left( \frac{N + R}{N - R} \right) = 0, \qquad (1.38)$$

which implies

$$\frac{p}{q} = \frac{(N + R_{max})}{(N - R_{max})}, \qquad (1.39)$$

which we can rearrange to get the value $R_{max}$ that maximizes $P$ as

$$R_{max} = (p - q)N, \qquad (1.40)$$

where we used $p + q = 1$. The value of $R$ that maximizes the probability is identical to the mean for the binomial distribution, which we found when determining $\langle R \rangle$ for a random walk. There is no guarantee that the mean and most probable value will coincide for a general probability distribution.

Having found the value of $R$ that maximizes the probability distribution, we can investigate the form of the distribution near the maximum by performing a Taylor expansion about $R_{\text{max}} = \langle R \rangle$:

$$\ln P(N, R) = \ln P(N, R_{\text{max}}) + (R - R_{\text{max}}) \left. \frac{\partial \ln P(N, R)}{\partial R} \right|_{R=R_{\text{max}}}$$

$$+ \frac{1}{2}(R - R_{\text{max}})^2 \left. \frac{\partial^2 \ln P(N, R)}{\partial R^2} \right|_{R=R_{\text{max}}} + \cdots . \qquad (1.41)$$

We know that the first derivative term vanishes because we demanded this in order to find the maximum, so we now evaluate

$$\left. \frac{\partial^2 \ln P(N, R)}{\partial R^2} \right|_{R=R_{\text{max}}} \simeq \left. \frac{\partial}{\partial R} \left[ \frac{1}{2} \ln \frac{p}{q} - \frac{1}{2} \ln(N + R) + \frac{1}{2} \ln(N - R) \right] \right|_{R=R_{\text{max}}}$$

$$= -\frac{1}{2} \left[ \frac{1}{N + R_{\text{max}}} + \frac{1}{N - R_{\text{max}}} \right]$$

$$= -\frac{1}{2N} \left[ \frac{1}{2p} + \frac{1}{2q} \right]$$

$$= -\frac{1}{4Npq}, \qquad (1.42)$$

where we used the expression $R_{\text{max}} = (p - q)N$ and $p + q = 1$ to write $N + R_{\text{max}} = 2pN$ and $N - R_{\text{max}} = 2qN$. Note that the second derivative is negative, confirming that we have a maximum of the probability distribution. Thus we have

$$\ln P(N, R) \simeq \ln P(N, R_{\text{max}}) - \frac{(R - R_{\text{max}})^2}{2\sigma_R^2} + \cdots , \qquad (1.43)$$

recalling that we found above that $\sigma_R^2 = 4Npq$.

We might worry about whether higher-order terms will turn out to be more important than the leading terms. To assuage these concerns, we can see that when we take another derivative of $\ln P$ with respect to $R$, we will get a term with $N^2$ in the denominator, so the third-order term will be of order $R^3/N^2$, which is much smaller than the second-order term, which is of order $R^2/N$ when $N$ is large. A more careful analysis finds that the condition that higher-order terms are unimportant is that $Npq \gg 1$, which will be true for large $N$ unless one of $p$ or $q$ is vanishingly small.

We now determine the value of $P$ at $R_{\text{max}}$ by calculating the term

$$\ln P(N, R_{\text{max}}) = \ln N! - \ln \left\{ \left[ \frac{1}{2}(N + R_{\text{max}}) \right]! \right\} - \ln \left\{ \left[ \frac{1}{2}(N - R_{\text{max}}) \right]! \right\}$$

$$+ \frac{1}{2}(N + R_{\text{max}}) \ln p + \frac{1}{2}(N - R_{\text{max}}) \ln q$$

$$\simeq \left( N + \frac{1}{2} \right) \ln(N) - N + \frac{1}{2} \ln(2\pi) - \frac{1}{2}(2pN + 1) \ln(pN)$$

$$+ pN - \frac{1}{2} \ln(2\pi) - \frac{1}{2}(2qN + 1) \ln(qN)$$

$$+ qN - \frac{1}{2} \ln(2\pi) + pN \ln p + qN \ln q$$

$$= -\frac{1}{2} \ln(2\pi pqN), \tag{1.44}$$

where we skipped a few lines of tedious algebra. Putting our results together we have, in the large-$N$ limit,

$$P(N, R) \simeq \frac{1}{\sqrt{2\pi pqN}} \exp\left\{ -\frac{(R - R_{\max})^2}{2\sigma_R^2} \right\}, \tag{1.45}$$

which we can see is a Gaussian distribution with a relative width that scales as $\sigma_R / R_{\max} \propto 1/\sqrt{N}$, which becomes infinitely sharp in the large-$N$ limit.

To check that this is normalized, we need to perform a sum over $R$. Assuming that we can treat $P(N, R)$ as continuous in $R$ (which will be a good approximation if the highly probable values of $R$ are much less than $N$), we can convert the sum to an integral by recalling that $R$ jumps in steps of two, so we need to calculate[2]

$$\int_{-\infty}^{\infty} \frac{dR}{2} P(N, R) = \frac{1}{2} \int_{-\infty}^{\infty} dR \frac{1}{\sqrt{2\pi pqN}} \exp\left\{ -\frac{(R - R_{\max})^2}{8pqN} \right\} = 1.$$

We can obtain a probability density $\mathcal{P}(R)$ for $R$ by rewriting

$$\int_{-\infty}^{\infty} \frac{dR}{2} P(N, R) = \int_{-\infty}^{\infty} dR \, \mathcal{P}(R),$$

where

$$\mathcal{P}(R) = \frac{1}{\sqrt{2\pi\sigma_R^2}} \exp\left\{ -\frac{(R - R_{\max})^2}{2\sigma_R^2} \right\}.$$

As a quick check, we can consider the case where $p = q = \frac{1}{2}$, corresponding to the coin toss example, in which case $R_{\max} = 0$, and we can ask what the probability of tossing equal numbers of heads and tails is out of 100 trials. Our expression above gives

$$P(N, 0) = \sqrt{\frac{2}{\pi N}}, \tag{1.46}$$

which for $N = 100$ is $P(100, 0) \sim 0.08$, as we found in Eq. (1.10). Gaussians of the form determined in Eq. (1.45) are illustrated in Fig. 1.5 for several different values of $N$ and $p \neq q$.

Hence we have shown that the large-$N$ limit of a binomial distribution is a Gaussian distribution. This is encouraging, because Gaussian distributions can often be easier to work with than the original distribution. This example also illustrates that while we may not be able to predict where an individual random walker ends up on the line, we can make precise predictions about the average behaviour if we have an ensemble of walkers.

---

[2] We review Gaussian integrals in Appendix A.

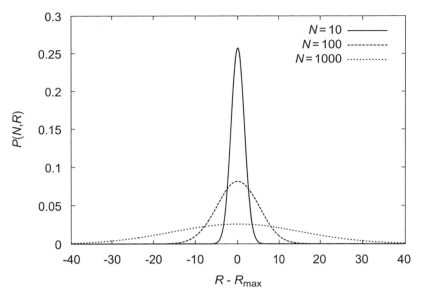

Fig. 1.5 Gaussian distributions $P(N, R)$ for $N = 10, 100, 1000$, with $p = 0.6, q = 0.4$.

## 1.2.6 Central Limit Theorem

The observation that the limiting form of the binomial distribution is a Gaussian distribution leads naturally to the question of whether this is specific to the binomial distribution or an example of more general behaviour. The answer turns out to be the latter, which can be formalized in the **central limit theorem**, which, for a collection of $N$ random variables, we state as:

> *The probability density of a collection of N independent random variables with finite variance, summed together, is a Gaussian distribution in the limit that $N \rightarrow \infty$.*

We will not prove this result, but it is worth spending a moment clarifying the content of this theorem. If we have $N$ random variables $s_i$ that can take on multiple different values (this could be more than the two-value case we considered above, e.g. six different possibilities when rolling a die) and we form their sum

$$S = \sum_{i=1}^{N} s_i, \tag{1.47}$$

then in the limit that $N \rightarrow \infty$ the probability distribution for the sum $S$ is

$$p(S) = \frac{1}{\sqrt{2\pi\sigma_S^2}} \exp\left\{-\frac{(S - \bar{S})^2}{2\sigma_S^2}\right\}, \tag{1.48}$$

where

$$\bar{S} = N\bar{s}; \qquad \sigma_S^2 = N\sigma_s^2, \tag{1.49}$$

with $\sigma_s^2 = \overline{s^2} - \bar{s}^2$. The Gaussian form relies on $S$ not being too far from the mean value of $S, \bar{S}$, and on the existence of the variance $\sigma_s^2$. This second point is very important as there are probability distributions that do not have a finite variance – the Lorentzian distribution is a well-known example.

The ramifications of this result are quite profound. First, the theorem implies that in many cases the details of the terms in the sum are not important. If we are interested in a probability distribution for macroscopic properties averaged over microscopic constituents, the sum will have a Gaussian probability distribution. This makes it not so surprising that Gaussians arise in nature in widely varying circumstances. Second, the theorem indicates that systems with large numbers of constituents will have well-defined properties, since the relative width of the distribution scales as

$$\frac{\sigma_S}{\bar{S}} \propto \frac{\sqrt{N}}{N} = \frac{1}{\sqrt{N}}, \tag{1.50}$$

which vanishes as $N \to \infty$. This property is one of the key features that allows a statistical mechanical description of nature to be effective.

## 1.3  Microstates and Macrostates

An important concept in statistical mechanics is that of the *state* of a system. In any physical system, there are a set of dynamical variables that characterize that particular system. We will mostly be concerned with *microstates* and *macrostates*. In a microstate, every dynamical variable has a definite value. Examples include:

  (i) the position and momenta of all the atoms in a classical gas;
 (ii) the quantum numbers of a set of simple harmonic oscillators;
(iii) the spins and positions of atoms in a crystal lattice.

A macrostate is the set of microstates with a particular energy $E$, which satisfy any other constraints on the system, such as fixed volume $V$ or particle number $N$. This set of microstates that constitute a macrostate are macroscopically indistinguishable.

In a classical system, the set of all permissable microstates (i.e. those with allowed values of $E$, $V$ and $N$) forms phase space. The most familiar example of phase space is the $6N$-dimensional space of $N$ classical particles in three-dimensional space, for which there are $3N$ position components $q_i$ and $3N$ momentum components $p_i$, where $i$ runs from 1 to $3N$. The microstates of the classical particle system are points in phase space $(p_i, q_i)$ represented by $6N$-dimensional vectors.

In a quantum mechanical system, microstates correspond to energy eigenstates (i.e. solutions of the time-independent Schrödinger equation) and the set of all microstates with the same energy $E$ forms a subspace of the Hilbert space for the system. An example of a collection of microstates is a set of degenerate levels for a particle in a three-dimensional box. These concepts of microstates and macrostates are most easily illustrated with an example.

### 1.3.1  Example: Non-interacting Spins in a Solid

A feature of quantum mechanics is that angular momentum is quantized. This leads to particles such as an electron having an intrinsic magnetic moment that is referred to as spin. These spins give rise to the magnetic properties of materials. For example, permanent magnetism, such as ferromagnetism in iron, arises from the spontaneous alignment of electron spins once iron is cooled below a special temperature, the Curie temperature, which is 1043 K (we will discuss this in more detail in Chapter 10). In ferromagnets such as iron we need to take into account interactions between spins in order to understand the magnetism. However, there are also materials (e.g. paramagnetic salts) in which there are atomic magnetic moments that are essentially independent of each other and hence can be viewed as a collection of magnetic dipoles that can be aligned by a magnetic field. For a single spin in a magnetic field with magnitude $B$ applied parallel to the $z$-axis, the energy takes the form

$$u = -m_z B, \tag{1.51}$$

where $m_z = \pm\mu$ is the magnetic moment of the spin, which we take to be in one of two possible states, up ($\uparrow$) corresponding to $m_z = \mu$, or down ($\downarrow$) corresponding to $m_z = -\mu$.

If we assume that the spins are located on $N$ distinct sites, and that each spin can either point up or down, then this example is just like the random walk or the coin toss experiment. A microstate of the system is a specification of the spin on each of the $N$ sites (e.g. see Fig. 1.6). This implies that there are $2^N$ possible microstates, and from the binomial expansion we see that if there are $N_\uparrow$ up spins and $N_\downarrow$ down spins, with

$$N_\uparrow + N_\downarrow = N, \tag{1.52}$$

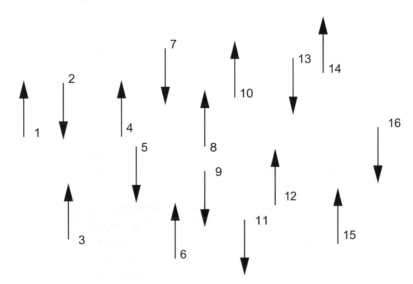

**Fig. 1.6**   A collection of non-interacting spins – the sites are distinguishable (numbered) but the spins are not.

there are

$$\frac{N!}{N_\uparrow! \, N_\downarrow!}$$

different microstates with the same macroscopic value of $N_\uparrow$.

We can also determine this from a combinatorial point of view. Imagine we take spins and put them sequentially at $N$ different sites. There are $N$ choices for the location of the first spin, $N - 1$ choices for the location of the second spin, etc., leaving $N(N - 1)(N - 2) \cdots 1 = N!$ ways to arrange the spins on the sites. If we restrict ourselves to up spins, there are similarly $N_\uparrow!$ ways to arrange these on the $N_\uparrow$ sites with up spins. The same logic holds for down spins. As the spins are identical, we need to divide $N!$ by $N_\uparrow! \, N_\downarrow!$ to determine the number of *distinguishable* arrangements of spins:

$$\Omega(N, N_\uparrow) = \frac{N!}{N_\uparrow! \, N_\downarrow!} = \binom{N}{N_\uparrow} = \frac{N!}{N_\uparrow! \, (N - N_\uparrow)!}, \tag{1.53}$$

where we have introduced the **multiplicity function** $\Omega(N, N_\uparrow)$ (also known as the statistical weight), which counts the number of microstates that have the same value of $N_\uparrow$ when there are $N$ spins. More generally, we shall use $\Omega$ to denote the multiplicity function in any situation where we are counting the number of microstates.

We can define

$$M = N_\uparrow - N_\downarrow,$$

which is the overall magnetization of the collection of spins in units of $\mu$. We can note that this is related to the total energy by

$$U = -\mu N_\uparrow B + \mu N_\downarrow B = -\mu M B. \tag{1.54}$$

We also have

$$N_\uparrow = \frac{1}{2}(N + M), \qquad N_\downarrow = \frac{1}{2}(N - M), \tag{1.55}$$

so we can write the multiplicity function as[3]

$$\Omega(N, M) = \frac{N!}{\left[\frac{1}{2}(N + M)\right]! \left[\frac{1}{2}(N - M)\right]!}. \tag{1.56}$$

Now, the probability of finding the spin system in any particular microstate is $p^{N_\uparrow} q^{N_\downarrow}$, where $p$ and $q$ are the probabilities for an individual spin to be up or down, and this gives $1/2^N$ if $p = q = 1/2$.

Hence the probability of finding the spin system in any particular macrostate (that is with $N_\uparrow$ spins pointing up and $N_\downarrow$ spins pointing down, without concern as to whether the spin is up or down on any particular site) will be the multiplicity function times the probability of having a microstate with $N_\uparrow$ spins pointing up and $N_\downarrow$ spins pointing down, i.e.

$$P(N, M) = \Omega(N, M) p^{\frac{1}{2}(N+M)} q^{\frac{1}{2}(N-M)}. \tag{1.57}$$

---

[3] $M$ plays the same role in this problem that $R$ plays in the random walk.

This is exactly the same probability distribution that we encountered in the random walk, and hence we already know several of its properties. However, we will first focus on some other aspects of this example, and use them to introduce the concept of entropy.

## 1.4 Information, Ignorance and Entropy

In the non-interacting spin example above, we can classify different macrostates by their magnetization $M$ (which is equivalent to the energy $U$). The multiplicity function $\Omega$ tells us the number of microstates in a given macrostate. The reason we want to know this is usually that we cannot determine the microstate of a system via a macroscopic measurement; we are only able to determine a macroscopic quantity like $M$.

Expanding on this idea, if all the spins are up, i.e. $M = N$, then there is only one microstate permitted in the corresponding macrostate, so there is no uncertainty as to which microstate the system is in. However, if $N - 1$ spins are up and 1 is down, i.e. $M = N - 2$, then there are $N$ different choices we can make for the location of the down spin and so there are $\Omega(N, N - 2) = N$ microstates corresponding to this macrostate. In contrast, if we know that the second spin is down, then the system is in a definite microstate. If there are 2 down spins, then there are $\Omega(N, N - 4) = \frac{1}{2}N(N - 1)$ microstates, and so on.

We notice from the above examples that if we measure a particular value of $M$, then we have more ignorance of which microstate the system is in when 2 spins are down than when 1 spin is down, or when 0 spins are down. To generalize from this example, we would like to be able to make comparisons between different macrostates to know how much ignorance we have of the state of the system for given macroscopic parameters. This leads us to a more precise definition of entropy in terms of our ignorance of microstates or the "uncertainty" or "amount of information" in a given probabilistic outcome, or collection of microstates.

Introduce a function $\sigma$ (we will later identify $\sigma$ with the entropy) which quantifies our ignorance of the microstate in which the system can be found. We will start our discussion here by assuming that

*All of the microstates in a given macrostate are equally likely.*

It is not immediately obvious that this should hold, but without additional information, we have no reason to believe that any one microstate with the same macroscopic properties should be any more probable than any other. This is not a justification, but is at least a plausibility argument. The best justification is that the theory based on this assumption has been tested very thoroughly and has had very wide success. We will re-examine this assumption when we discuss the microcanonical ensemble in Chapter 2. Note that there should be no expectation that microstates from different macrostates have the same likelihood. Once we assume all the microstates in a macrostate are equally likely, then $\sigma$ should just be a function of the number of accessible microstates, given by the multiplicity function $\Omega$, hence $\sigma = \sigma(\Omega)$. There are a number of constraints on this function that we can deduce:

(A) If $\Omega = 1$, the microstate is determined and we have no ignorance, so

$$\sigma(1) = 0. \tag{1.58}$$

(B) If we compare two macrostates 1 and 2 with multiplicity functions $\Omega_1$ and $\Omega_2$, then if $\Omega_1 > \Omega_2$ (i.e. we have more ignorance about macrostate 1 than macrostate 2):

$$\sigma(\Omega_1) > \sigma(\Omega_2). \tag{1.59}$$

(C) Suppose we have two independent systems (e.g. two independent sets of spins, with given magnetizations $M_1$ and $M_2$) so the total multiplicity function for the joint state of the two systems is $\Omega = \Omega_1\Omega_2$, then if the microstate associated with macrostate 1 is fixed, i.e. $\Omega_1 = 1$, we retain all of our ignorance of macrostate 2.

This suggests

$$\sigma(\Omega) = \sigma(\Omega_1\Omega_2) = \sigma(\Omega_1) + \sigma(\Omega_2). \tag{1.60}$$

Note that if we had taken $\sigma(\Omega_1\Omega_2) = \sigma(\Omega_1)\sigma(\Omega_2)$, then we would get $\sigma(\Omega) = 0$ if $\sigma(\Omega_1) = 0$, regardless of $\Omega_2$, which would contradict our condition (B). Hence we would like a functional form for $\sigma$ that satisfies the following:

$$\sigma(1) = 0, \tag{1.61}$$

$$\sigma(xy) = \sigma(x) + \sigma(y), \tag{1.62}$$

$$\sigma(x) > \sigma(y) \quad \text{if} \quad x > y. \tag{1.63}$$

To find such a form, take derivatives of Eq. (1.62) with respect to $x$ and $y$:

$$\frac{\partial\sigma(xy)}{\partial x} = \frac{d\sigma(x)}{dx}, \tag{1.64}$$

$$\frac{\partial\sigma(xy)}{\partial y} = \frac{d\sigma(y)}{dy}, \tag{1.65}$$

and if we let $z = xy$, then using the chain rule

$$\frac{d\sigma(z)}{dz} = \frac{dx}{dz}\frac{\partial\sigma(z)}{\partial x} = \frac{1}{y}\frac{\partial\sigma(z)}{\partial x}, \tag{1.66}$$

which we can write as

$$\frac{\partial\sigma(z)}{\partial x} = y\frac{d\sigma(z)}{dz}. \tag{1.67}$$

Similarly

$$\frac{\partial\sigma(z)}{\partial y} = x\frac{d\sigma(z)}{dz}, \tag{1.68}$$

and so when we combine Eqs (1.64), (1.65), (1.67) and (1.68) we get

$$x\frac{d\sigma(x)}{dx} = y\frac{d\sigma(y)}{dy} = z\frac{d\sigma(z)}{dz} = k, \tag{1.69}$$

where we note that each term is a function of independent variables and hence must be equal to a constant $k$. We can thus integrate Eq. (1.69) with respect to $x$ to get

$$\sigma(x) = k \ln(x) + c, \tag{1.70}$$

where $c$ is a constant, and $\sigma(1) = 0$ implies $c = 0$, so

$$\sigma(\Omega) = k \ln(\Omega). \tag{1.71}$$

If we set the constant $k$ to be equal to Boltzmann's constant $k_B = 1.3807 \times 10^{-23}$ J K$^{-1}$ then we obtain Boltzmann's equation for the entropy:

$$S = k_B \ln \Omega. \tag{1.72}$$

We can view this as our "ignorance" associated with a particular macrostate, or alternatively the "information content" of a macrostate.

## 1.5 Summary

In systems with many macroscopic degrees of freedom it is not possible to describe the system in terms of all of the microscopic degrees of freedom, and hence a statistical description is more appropriate. In the limit of large numbers of degrees of freedom, the central limit theorem states that sums of independent random variables have a Gaussian distribution, illustrated by the example of a random walk in one dimension.

The state of a physical system can be characterized either at the microscopic level – a microstate, or at the macroscopic level – a macrostate. Typically, many microstates may correspond to the same macrostate. We can quantify our ignorance as to which microstate the system is in for a given macrostate by the entropy, which, when all microstates within a given macrostate are equally likely, takes the form

$$S = k_B \ln \Omega. \tag{1.73}$$

## Problems

**1.1**   (a) We are told that

$$P(x) = \mathcal{N} x e^{-x}$$

is a probability distribution on $0 \le x < \infty$. Find the value of $\mathcal{N}$ so that $P(x)$ is properly normalized and then calculate $\langle x \rangle$ and $\langle x^2 \rangle$.

(b) Now consider the following probability distribution that depends on two variables, $x$ and $y$:

$$P(x, y) = \begin{cases} A(x^2 + y^2), & 0 < x < 1;\ 0 < y < 1, \\ 0, & \text{elsewhere.} \end{cases}$$

Calculate the value of the normalization constant $A$ and then find $\langle x \rangle$, $\langle y \rangle$ and $\langle x^2 + y^2 \rangle$.

**1.2** For a uniform probability distribution on the surface of a sphere,

$$P(\theta, \phi) = \frac{1}{4\pi},$$

calculate the following averages:

(a) $\langle \cos \phi \rangle$;
(b) $\langle \cos \theta \cos \phi \rangle$;
(c) $\langle \cos^2 \theta \cos^2 \phi \rangle$;
(d) $\langle \theta^2 \phi^2 \rangle$.

**1.3** Consider a simple harmonic oscillator for which the displacement is

$$x(t) = A \cos(\omega t).$$

(a) Show that the probability that the oscillator is found between $x$ and $x + dx$ is given by

$$P(x) = \begin{cases} \dfrac{1}{\pi \sqrt{A^2 - x^2}}, & |x| \leq A, \\ 0, & |x| > A. \end{cases}$$

*Hint:* consider the time $dt$ that the oscillator spends in the interval $[x, x + dx]$.

(b) Check that the probability you found in part (a) is properly normalized and find the averages $\langle x \rangle$ and $\langle x^2 \rangle$ for the simple harmonic oscillator.

**1.4** In the *continuum limit* of long length and time scales, random walks may be described using the diffusion equation

$$\frac{\partial \rho}{\partial t} = D\nabla^2 \rho,$$

where $\rho(x, t)$ is the probability density for a walker. We will derive this result by considering a general random walk in one dimension in which there are no correlations between indvidual steps in the walk. At each time $t$, the position at time $t + \delta t$ is given by

$$x(t + \delta t) = x(t) + z(t),$$

where $z$ is chosen from the probability distribution $\chi(z)$, which is a function of $z$ with mean zero and standard deviation $a$, and $\delta t$ is taken to be fixed.

(a) Write down $\chi(z)$ for an unbiased random walk with constant step size in one dimension.

(b) Writing $\chi$ as a function of $z$, write down equations for the zeroth, first and second moments of $\chi$.

(c) Show that

$$\rho(x, t + \delta t) = \int_{-\infty}^{\infty} dz\, \rho(x - z, t)\chi(z).$$

(d) Assume that the typical step size in space is small on the scale on which $\rho$ varies and Taylor expand $\rho$ to obtain the diffusion equation. Specify the diffusion constant $D$.

**1.5**    Consider a collection of 70 non-interacting spins that may point either up or down. How many microstates are there in this system? How can we label the macrostates in this system? If each microstate is equally probable, what is the most likely macrostate to find the system in, and what is the probability that we will find the system in this configuration?

**1.6**    Bayes' theorem can be stated as

$$P(A|B) = \frac{P(B|A)P(A)}{P(B)},$$

where $P(A|B)$ is the probability of $A$ given that we know $B$ has occurred. $P(B|A)$ is the probability of $B$ given that $A$ is true and $P(A)$ is the prior, which reflects our best estimate of the probability of $A$.

Suppose we have a choice of five different dice, four of which are fair and one of which is known to be loaded so that there is a 1/3 chance that a six will be rolled. We choose one die then roll five consecutive sixes with the die we chose. Note that the probability of rolling five consecutive sixes, which we denote as $P(5 \times 6)$, is

$$P(5 \times 6) = P(5 \times 6|\text{fair})P(\text{fair}) + P(5 \times 6|\text{not fair})P(\text{not fair}),$$

where $P(\text{fair})$ is the probability that we picked a fair die and $P(\text{not fair})$ is the probability that we picked the unfair die. Given the above information, determine the probability that we picked a fair die.

**1.7**    Certain marine bacteria can explore their surroundings through a two-step process. They "run" in a straight line at speeds of order 100 μm s$^{-1}$ for around 100 ms, and then reorient themselves randomly over a much shorter time period. Estimate the volume of seawater that one of these bacteria can explore in 10 minutes.

**1.8**    Consider a biased random walk in one dimension with probability $p$ for a step to the right and probability $q$ for a step to the left. Let the time step be $\tau_0$, so that the number of steps $N$ is related to the time $t$ by $N = t/\tau_0$. What is the characteristic time scale $\tau_X$ for the crossover from diffusive to ballistic motion?

**1.9**    A polymer consists of $N$ molecular units, each of length $a$, and is in a state of thermodynamic equilibrium at temperature $T$. The units are joined so as to permit a free rotation in any direction about the joints.

(a) By treating the freely jointed chain in the absence of force as a random walk of monomers in three dimensions, argue that in the limit of large $N$, the probability that the end-to-end displacement of the chain is $\mathbf{R}$ is given by

$$P(\mathbf{R}, N) = \left(\frac{3}{2\pi N a^2}\right)^{\frac{3}{2}} e^{-\frac{3R^2}{2Na^2}}.$$

(b) Suppose a small force $\mathbf{F} = F\hat{\mathbf{z}}$ is applied to the chain which leads to a small displacement $\mathbf{R} \rightarrow \mathbf{R} + \delta\mathbf{R}$. Assuming that the deformation conserves energy and that the change is reversible, show that we may view the polymer as a spring with a spring constant

$$k = \frac{3k_B T}{Na^2}.$$

*Hint:* it may be useful to recall that $dU = dW + TdS$.

**1.10** Perform a numerical simulation of a random walk to generate a trace similar to those shown in Fig. 1.4. Choose the probability of a move to the right to be $p$, and let the step size for the walker be fixed at 1. Take the initial position of the walker to be $x = 0$.

(a) At each step of the walk, generate a random number $r$ between 0 and 1. If $r \leq p$ then $x \rightarrow x + 1$, and if $r > p$ then $x \rightarrow x - 1$. After $N$ steps (try several different values, e.g. $N = 100, 1000, 10\,000$) you will have a random walk. Record $x$ and $x^2$ for this walk.

(b) Now repeat the $N$-step walk $M$ times (where $M$ should be of order at least 100). Calculate

$$\langle x \rangle_N = \frac{1}{M} \sum_{m=1}^{M} x_{m,N}$$

and

$$\left\langle x^2 \right\rangle_N = \frac{1}{M} \sum_{m=1}^{M} x_{m,N}^2,$$

where $x_{m,N}$ is the position of the walker in the $m$th random walk after $N$ steps, $\langle x \rangle_N$ is the mean displacement after $N$ steps and $\left\langle x^2 \right\rangle_N$ is the mean square displacement after $N$ steps. Take data for at least 15 different values of $N$ between $N = 10$ and $N = 10\,000$ and compare your results for $\langle x \rangle_N$ and $\left\langle x^2 \right\rangle_N$ to the expressions in Eqs (1.27) and (1.29). Additionally, calculate the variance

$$\sigma_N^2 = \left\langle x^2 \right\rangle_N - \langle x \rangle_N^2,$$

and compare your result to Eq. (1.30).

# 2     The Microcanonical Ensemble

There are three important statistical ensembles used widely in physics, each of which describe systems with different constraints. These are known as the microcanonical, the canonical and the grand canonical ensembles. The ensembles are distinguished by their interaction with the rest of the world: the microcanonical ensemble is for isolated systems, which do not exchange energy or particles with the rest of the world; the canonical ensemble deals with closed systems, which exchange energy but not particles with the rest of the world; and the grand canonical ensemble is for open systems, which can exchange energy and particles with the rest of the world. For a given situation of interest, one or more of these ensembles may be appropriate for determining the relevant physical properties.

Josiah Willard Gibbs introduced the idea of ensembles as a way to give concrete meaning to the probabilities of states that we discussed in Chapter 1. An ensemble can be viewed as a very large number of copies of the system of interest, each prepared in an identical way. If we have very many of these copies of the system, then the probability of finding the system in any particular microstate will be given by the fraction of members of the ensemble which are in that microstate. We will start by discussing the microcanonical ensemble.

In the microcanonical ensemble, the energy of each member of the ensemble is fixed (within a very narrow range), so we can think of an ensemble of thermally isolated systems, each with the same energy $U$.[1] We also assume (if relevant) that the number of particles, $N$, and the volume, $V$, are fixed for each member of the ensemble. In practice, apart from some specially prepared systems of ultra-cold atoms which have almost no interaction with the rest of the world, most real systems have some level of energy exchange with their environment and a truly thermally isolated system is an idealization. When we calculate quantities in the microcanonical ensemble (or other ensembles), we are performing an equilibrium average over all microstates that are consistent with the constraints (in this case constant $U$, $N$ and $V$) on the system, as illustrated in Fig. 2.1 for example. In this chapter we will connect the microscopic statistical properties of an isolated thermal system to its thermodynamic properties.

---

[1]   If we are thinking about microstates in a quantum system as energy eigenstates, then the picture is of the system having multiple accessible microstates, all close in energy to each other, rather than the system being in a single eigenstate with energy $U$.

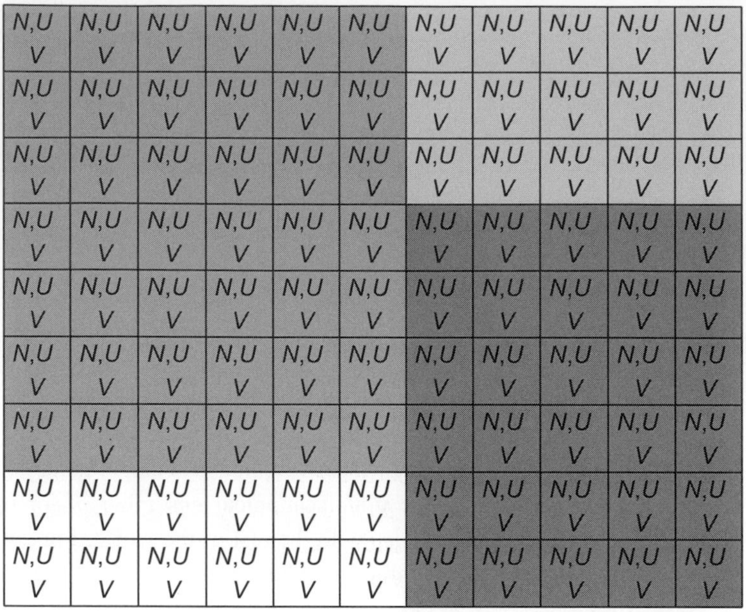

| N,U V | N,U V | N,U V | N,U V | N,U V | N,U V | N,U V | N,U V | N,U V | N,U V | N,U V |
|-------|-------|-------|-------|-------|-------|-------|-------|-------|-------|-------|
| N,U V | N,U V | N,U V | N,U V | N,U V | N,U V | N,U V | N,U V | N,U V | N,U V | N,U V |
| N,U V | N,U V | N,U V | N,U V | N,U V | N,U V | N,U V | N,U V | N,U V | N,U V | N,U V |
| N,U V | N,U V | N,U V | N,U V | N,U V | N,U V | N,U V | N,U V | N,U V | N,U V | N,U V |
| N,U V | N,U V | N,U V | N,U V | N,U V | N,U V | N,U V | N,U V | N,U V | N,U V | N,U V |
| N,U V | N,U V | N,U V | N,U V | N,U V | N,U V | N,U V | N,U V | N,U V | N,U V | N,U V |
| N,U V | N,U V | N,U V | N,U V | N,U V | N,U V | N,U V | N,U V | N,U V | N,U V | N,U V |
| N,U V | N,U V | N,U V | N,U V | N,U V | N,U V | N,U V | N,U V | N,U V | N,U V | N,U V |
| N,U V | N,U V | N,U V | N,U V | N,U V | N,U V | N,U V | N,U V | N,U V | N,U V | N,U V |

**Fig. 2.1**   Microcanonical ensemble: each member of the ensemble has identical energy $U$, particle number $N$ and volume $V$, but may be in a different microstate – distinct microstates are indicated by different background shading.

## 2.1 Thermal Contact

The concept of thermal equilibrium is central to thermodynamics and will be very important in our discussions of statistical mechanics. Generically, equilibrium is the state that a system reaches at long times which is independent of the details of initial conditions. This implies that equilibration is an irreversible process. If we view a state in terms of probabilities, then for given constraints, equilibrium will occur when the number of possible microstates is maximized, since there will be more ways to achieve this (and hence higher probability) than when the number of microstates is not maximal.

In order to gain insight into microcanonical systems, suppose we bring two isolated systems with fixed volumes into thermal contact, which will allow flow of energy between them (see Fig. 2.2). We will assume that the two systems have energies $U_1$ and $U_2$, respectively and $N_1$ and $N_2$ particles, respectively (we use particle in a broad sense – this could correspond to spins for instance), so the total energy of the combined system is

$$U = U_1 + U_2,$$

and there are $N = N_1 + N_2$ particles. We expect that when we bring the systems into contact they will exchange energy until they reach thermal equilibrium. The exact form of that equilibrium will depend on the details of the different sizes and initial energies of the systems, but we will see below that we can make quite concrete statements about the relationship between statistical and thermal properties of equilibrium systems.

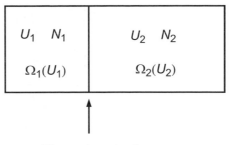

Thermal contact

**Fig. 2.2** Thermal contact between two initially isolated systems. System 1 has energy $U_1$, particle number $N_1$ and multiplicity function $\Omega_1(U_1)$. System 2 has energy $U_2$, particle number $N_2$ and multiplicity function $\Omega_2(U_2)$.

We assume that there is a barrier between the systems that allows for energy flow, but not for the flow of particles, so $N_1$ and $N_2$ remain fixed. The total energy $U$ will be conserved, but $U_1$ and $U_2$ may change with time. The number of accessible microstates in each system is $\Omega_1(U_1)$ and $\Omega_2(U_2)$, respectively, so the total multiplicity function for the system, which will depend on both $U$ and $U_1$ (or equivalently $U_2$, since $U_1$ and $U_2$ are not independent once $U$ is known) is

$$\Omega(N, U, U_1) = \Omega_1(U_1)\Omega_2(U_2) = \Omega_1(U_1)\Omega_2(U - U_1), \tag{2.1}$$

where we have suppressed the labels $N_1$ and $N_2$ since they are fixed. If we assume that all microstates are equally likely (as we did in our derivation of the Boltzmann entropy), then the most probable values of $U_1$ and $U_2$ are those which maximize $\Omega(N, U, U_1)$. In the limit that $N_1, N_2 \gg 1$, we can use the central limit theorem (Section 1.2.6) to argue that $\Omega_1$ and $\Omega_2$ will have Gaussian forms, hence

$$\Omega(N, U, U_1) = \frac{1}{2\pi\sigma_1\sigma_2} \exp\left\{-\left[\frac{(U_1 - \hat{U}_1)^2}{2\sigma_1^2} + \frac{(U - U_1 - \hat{U}_2)^2}{2\sigma_2^2}\right]\right\}, \tag{2.2}$$

where $\hat{U}_1$ and $\hat{U}_2 = U - \hat{U}_1$ are the most probable values of $U_1$ and $U_2$. These values scale with $N_1$ and $N_2$, respectively and $\sigma_1^2$ and $\sigma_2^2$ are the variances for $\Omega_1$ and $\Omega_2$ which also scale with $N_1$ and $N_2$. As we saw from the central limit theorem, the relative widths of the Gaussians for $\Omega_1$ and $\Omega_2$ will scale as $1/\sqrt{N_1}$ and $1/\sqrt{N_2}$, so that the most probable state of the entire system (the one in which $U_1 = \hat{U}_1$ and $U_2 = \hat{U}_2$) is overwhelmingly more likely than any other state. For a macroscopic number of particles $N_{1,2} \sim 10^{23}$, this means that we will essentially never see fluctuations that lead $U_1$ or $U_2$ to deviate from their most probable values.

As emphasized above, the most probable values of $U_1$ and $U_2$ are those that maximize $\Omega(N, U, U_1)$, i.e.

$$\left.\frac{\partial}{\partial U_1}\Omega(N, U, U_1)\right|_{U_1=\hat{U}_1} = 0. \tag{2.3}$$

Maximizing $\Omega$ is equivalent to maximizing $\ln \Omega$, i.e.

$$\frac{\partial}{\partial U_1} \ln \Omega \bigg|_{U_1 = \hat{U}_1} = 0. \tag{2.4}$$

In Chapter 1 we identified $\ln \Omega = S/k_B$, where $S$ is the entropy. Thus, this corresponds to the result from thermodynamics that the entropy is maximum for an isolated system at equilibrium, so the most likely state of the system is the one with the maximum entropy.

We can express the condition for the maximum in $\Omega(N, U, U_1)$ as

$$\left(\frac{\partial \Omega_1(U_1)}{\partial U_1}\right)\bigg|_{U_1 = \hat{U}_1} \Omega_2(\hat{U}_2) + \Omega_1(\hat{U}_1) \left(\frac{\partial \Omega_2(U_2)}{\partial U_2}\right)\bigg|_{U_2 = \hat{U}_2} \frac{\partial U_2}{\partial U_1} = 0. \tag{2.5}$$

Now, $U_2 = U - U_1$ implies that

$$\frac{\partial U_2}{\partial U_1} = -1, \tag{2.6}$$

so we have

$$\frac{1}{\Omega_1} \frac{\partial \Omega_1}{\partial U_1} - \frac{1}{\Omega_2} \frac{\partial \Omega_2}{\partial U_2} = 0, \tag{2.7}$$

and then

$$\left(\frac{\partial \ln \Omega_1(U_1)}{\partial U_1}\right)\bigg|_{U_1 = \hat{U}_1} = \left(\frac{\partial \ln \Omega_2(U_2)}{\partial U_2}\right)\bigg|_{U_2 = \hat{U}_2}, \tag{2.8}$$

where we used

$$\frac{1}{y} \frac{dy}{dx} = \frac{d}{dx} \ln y. \tag{2.9}$$

Each side of Eq. (2.8) depends on independent variables, so both terms must be equal to a constant, say $\beta$. Hence we get at equilibrium that

$$\beta = \left(\frac{\partial \ln \Omega(U)}{\partial U}\right)_{U = \hat{U}}, \tag{2.10}$$

so if we identify $k_B \ln \Omega$ with the entropy $S$, then compatibility with thermodynamics requires $\beta = 1/k_B T$ and we have the relation

$$\frac{1}{T} = \frac{\partial S}{\partial U}\bigg|_{N,V}, \tag{2.11}$$

where $T$ is the temperature.

## 2.1.1 Heat Flow in the Approach to Equilibrium

In our discussion above, we have found that when we have equilibrium, the temperature $T$ is the same in both subsystems 1 and 2. The initial values of $U_1$ and $U_2$ (before the two subsystems are placed in thermal contact) are not in general equal to their equilibrium values $\hat{U}_1$ and $\hat{U}_2$, and our argument for the equality of the temperatures relied on the system being in equilibrium, so we may expect that there are temperatures $T_1$ and $T_2$

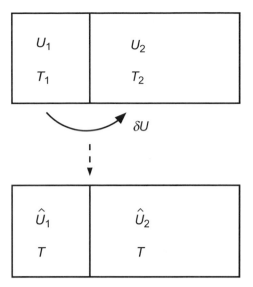

**Fig. 2.3** Heat flow in the equilibration of two subsystems in thermal contact. Energy $\delta U$ flows from system 1 to system 2, so that the energy of each system tends to its equilibrium value: $U_1 \to \hat{U}_1$ and $U_2 \to \hat{U}_2$ and the temperatures $T_1$ and $T_2$ both tend to the equilibrium temperature $T$.

associated with each subsystem before they are in contact. In order for the energies of each system to reach their equilibrium values, there must be a flow of energy $\delta U = U_1 - \hat{U}_1$ between the two systems.

As we can see from the dependence of the multiplicity functions $\Omega_1$ and $\Omega_2$ on $U_1$ and $U_2$, there will be a change in entropy

$$dS = \left(\frac{\partial S_1}{\partial U_1}\right) dU_1 + \left(\frac{\partial S_2}{\partial U_2}\right) dU_2, \tag{2.12}$$

and if we let the change in entropy $dS = \delta S$ when there is a change in energy $dU_1 = -\delta U$ as shown in Fig. 2.3, then

$$\delta S = \left(\frac{\partial S_1}{\partial U_1}\right)\Bigg|_{N_1} (-\delta U) + \left(\frac{\partial S_2}{\partial U_2}\right)\Bigg|_{N_2} (\delta U) = \left(-\frac{1}{T_1} + \frac{1}{T_2}\right) \delta U. \tag{2.13}$$

We argued that in equilibrium the energy changes to a value that maximizes the multiplicity function (i.e. the entropy), so $\delta S$ must be positive, and hence $T_1 > T_2$ if $\delta U > 0$. This implies that in the approach to equilibrium, energy flows from a hotter body to a colder body in order to maximize the total entropy, $S_{\text{total}}$, in accord with our expectations from thermodynamics.

## 2.1.2  Principle of Maximum Entropy

In the approach to equilibrium, we saw that there is a flow of heat such that the final entropy $S_{\text{final}}$ is greater than or equal to the initial entropy $S_{\text{initial}}$ and so

$$S_{\text{final}} \simeq k_B \ln \left(\Omega_1 \Omega_2\right)_{\text{max}} \geq S_{\text{initial}} = k_B \ln \left(\Omega_1 \Omega_2\right)_0, \tag{2.14}$$

which is an example of the principle that

*The entropy of a closed system tends to remain constant or increase when an internal constraint of the system is removed.*

Now, in theory, we could observe fluctuations in which the entropy decreases. However, when $N$ is large, this is overwhelmingly unlikely, so the principle of maximal entropy as stated above holds. In small systems, where $N$ is not large, there can be observable fluctuations in which entropy decreases, however on *average*, entropy will still increase. This is also known as the second law of thermodynamics.

In practice, the removal of constraints in a system can take many different forms – in essence removing constraints corresponds to making more states of the system accessible. An example might be having a container of gas with a partition that divides the container into two parts. Removing the barrier gives a larger volume for molecules to move in, increasing the number of accessible states, and hence the entropy.

When we consider that the underlying equations[2] for the evolution of the system in time are generally reversible, it is not obvious that the entropy should increase. The Poincaré recurrence theorem is a result that an isolated finite mechanical system will return to its initial state (or very close to its initial state) in a finite time (this time may be very long, but is not infinite). Near such a recurrence, one would expect the entropy to decrease. That we generally don't see such behaviour reflects that for macroscopic systems, recurrence times should be extremely long, much larger than the age of the universe. Ludwig Boltzmann[3] was aware of these issues and derived his celebrated "$H$-theorem" to show that interactions between particles tend to cause a quantity $H$ that he identified with the entropy to increase. In deriving this theorem, Boltzmann made the assumption of "molecular chaos" that the particles' motions are not correlated – this breaks time reversibility as interactions between the particles lead to their motion developing correlations as time increases, so that at long times it is not fully consistent to assume that particle motions are independent. While in principle the second law of thermodynamics is a statement about what should happen *most* of the time, in practice it is a statement about what happens *almost all* of the time.

### 2.1.3 Energy Resolution and Entropy

In introducing the microcanonical ensemble we considered the energy $U$ to be fixed within some narrow range, say $\delta U$. We can ask what effect this energy range has on the entropy. The multiplicity function (i.e. the total number of states with energy $U$) will be the product of the density of states at energy $U$, $g(U)$, and the width of the energy window:

$$\Omega(U) = g(U)\delta U,$$

---

[2] Either Hamilton's equations for classical systems or the Schrödinger equation for quantum systems.
[3] Boltzmann was one of the pioneers of statistical mechanics and formulated the relationship between the entropy and the multiplicity function, Eq. (1.72). He also made many contributions to kinetic theory and was particularly concerned with understanding how growth of the entropy occurs despite the time reversibility of the equations of motion.

from which we obtain

$$S(U) = k_B \ln \Omega(U) = k_B \ln g(U) + k_B \ln(\delta U). \tag{2.15}$$

Now, $\delta U$ is fixed, whereas the entropy is an extensive function, and scales with $N$, so the contribution to the entropy per particle from the term involving $\delta U$ scales as $k_B \ln(\delta U)/N$, which tends to zero as $N \rightarrow \infty$, demonstrating that the entropy is not very sensitive to the choice of $\delta U$.

## 2.2 Gibbs Entropy

In our derivation of the Boltzmann entropy in Chapter 1, we assumed that in equilibrium all microstates within a macrostate are equally likely. We can examine this assumption through the use of the microcanonical ensemble. Suppose that there are $M$ microstates which have energy $U$. For an ensemble with $\mathcal{N}$ members, for each microstate $i$, $\mathcal{N}_i$ members of the ensemble will be in that microstate, and so $\mathcal{N} = \sum_{i=1}^{M} \mathcal{N}_i$. We can determine the probability of each of the microstates by their frequencies of occurrence in the ensemble. The total number of distinct ways to place different members of the ensemble in the different microstates will be given by (the different members of the ensemble are taken to be distinguishable – we can imagine them on some lattice as in Fig. 2.1)

$$W(\mathcal{N}, U) = \frac{\mathcal{N}!}{\mathcal{N}_1! \, \mathcal{N}_2! \dots \mathcal{N}_M!}. \tag{2.16}$$

Assume that no member of the ensemble has a higher propensity to be in any one particular microstate than the others – this is different to assuming that all microstates have the same likelihood, we are just assuming that no member of the ensemble is biased more towards one microstate than any of the others. We can then determine the entropy of the entire ensemble as

$$S = k_B \ln W, \tag{2.17}$$

which we can evaluate with Stirling's approximation (keeping only the most important terms) as

$$\begin{aligned}
S &= k_B \left[ \mathcal{N} \ln \mathcal{N} - \sum_{i=1}^{M} \mathcal{N}_i \ln \mathcal{N}_i \right] \\
&= k_B \left[ \sum_{i=1}^{M} \mathcal{N}_i \ln \mathcal{N} - \sum_{i=1}^{M} \mathcal{N}_i \ln \mathcal{N}_i \right] \\
&= -k_B \mathcal{N} \sum_{i=1}^{M} \frac{\mathcal{N}_i}{\mathcal{N}} \ln \frac{\mathcal{N}_i}{\mathcal{N}},
\end{aligned} \tag{2.18}$$

and when we identify $\mathcal{N}_i/\mathcal{N} = p_i$, the probability of finding the system in microstate $i$, we can find that the average entropy for each member of the ensemble is given by

$$S_G = -k_B \sum_i p_i \ln p_i, \tag{2.19}$$

which is Gibbs' formula for the entropy. This generalizes Boltzmann's expression to the situation where not all microstates are equally likely, which is important for systems that are not in equilibrium or where not all states have the same energy, where the assumption of equal probability may not hold.

Usually we do not know the individual $p_i$, so one way for us to proceed is to use a strategy of *least bias*, by assuming that all microstates are equally likely, as we did in Chapter 1. If we revert to our earlier notation and refer to the total number of microstates as $\Omega$, then if all microstates are equally likely, the probability of finding the system in microstate $i$ will be $p_i = 1/\Omega$. This is actually equivalent to a strategy of *maximum uncertainty* in which we take the point of view that all things being equal we are more likely to be ignorant than not, so we should assume that our ignorance is maximal. This is also in accord with the requirement in thermodynamics that at fixed particle number and volume, the entropy should be maximal. We can follow this line of reasoning mathematically by maximizing

$$S_G - \lambda \left( \sum_i p_i - 1 \right),$$

where we introduced the Lagrange multiplier $\lambda$ to enforce the constraint that the sum of the $p_i$ must be 1. Now to maximize $S_G$ with respect to each of the $p_i$, we must have

$$\frac{\partial}{\partial p_i} \left[ -k_B \sum_j p_j \ln(p_j) - \lambda \left( \sum_j p_j - 1 \right) \right] = -k_B \ln p_i - k_B - \lambda = 0, \tag{2.20}$$

which we note is the same equation for each $p_i$, and hence all the $p_i$ must be equal, which gives us the same result as the strategy of least bias. With all the $p_i = 1/\Omega$, we obtain

$$S_G = -k_B \sum_i p_i \ln(p_i)$$

$$= -k_B \Omega \frac{1}{\Omega} \ln \left( \frac{1}{\Omega} \right)$$

$$= k_B \ln \Omega, \tag{2.21}$$

which is the form of the Boltzmann entropy introduced in Eq. (1.72).

## 2.3  Shannon Entropy

In Chapter 1 we introduced entropy as a measure of ignorance about the physical state of a system. The Shannon entropy is a closely related definition of entropy that arises in information theory. It is defined in the same way as the Gibbs entropy, Eq. (2.19), up to an overall scale factor:

$$S_{\text{Shannon}} = -k_s \sum_i p_i \ln p_i, \tag{2.22}$$

where $k_s = 1/\ln 2$. This formula measures information entropy in bits, which take either the value 0 or 1. To see that this is the case, consider a situation in which there are $N$ bits, in which case there are $\Omega = 2^N$ possible states. If all the states are equally probable, then all of the $p_i = 1/\Omega$ and

$$S_{\text{Shannon}} = \frac{\ln \Omega}{\ln 2} = \frac{\ln 2^N}{\ln 2} = N. \tag{2.23}$$

Naively we might think that when we know the microstate a system is in, i.e. our ignorance is zero, then this corresponds to having maximal information. The physical entropy is $S = 0$ in this case, as is the information entropy. Rather than measuring the information we have about a system, we can think of the information entropy as measuring the capacity of that system to communicate information. For instance, if we consider a spin microstate as a message, then when $S = 0$, it is because there is no capacity to convey information, because the message is already known. In contrast, if we have a spin system with the macroscopic property of zero magnetization, then there are many possible microstates consistent with our knowledge of the system, corresponding to many possible messages (one per microstate) that the spin system could be used to communicate.

We saw that we can view thermodynamic entropy in terms of information about a system. As discussed by Rolf Landauer, manipulating information also has thermodynamic implications. Suppose we have a computing device that stores $N$ bits of information and interacts with an environment which is at a temperature $T$. Landauer showed that erasing all of the stored information, e.g. by setting all the bits to 0, which is an irreversible process, has an energy cost. Bits are binary variables, like spins, so if we have $N$ bits, then similarly to spins, there are $2^N$ possible microstates, corresponding to a thermodynamic entropy of $S = N k_B \ln 2$ (Landauer, 1961). If we perform an operation in which we erase the memories, then the microstate is one in which all bits take the value 0, corresponding to an entropy of $S = 0$. Thus the change in entropy in resetting all of the bits is $\Delta S = -N k_B \ln 2$, or $-k_B \ln 2$ per bit. From the second law of thermodynamics,[4] the entropy change in any process must always be non-negative, so the entropy of the surroundings must increase by $k_B \ln 2$ per bit in order to compensate the lowering of entropy in the system. Hence the minimum heat that must be dissipated in the surroundings of the computational device is

$$T \Delta S = k_B T \ln 2 \tag{2.24}$$

per bit. Current computational devices are not close to reaching this limit.

---

[4] See Appendix B for a primer on thermal physics.

## 2.4  Example: Non-interacting Spins in a Solid

We return to the example of non-interacting spins that we discussed in Section 1.3.1 and recall that we noted that the magnetization $M$ in units of $\mu$ and the energy were related through

$$M = -\frac{U}{\mu B}, \tag{2.25}$$

so we may write the multiplicity function determined in Eq. (1.56) as a function of the energy:

$$\Omega(N,U) = \frac{N!}{\left[\frac{1}{2}\left(N - \frac{U}{\mu B}\right)\right]!\left[\frac{1}{2}\left(N + \frac{U}{\mu B}\right)\right]!}, \tag{2.26}$$

and hence the Boltzmann entropy is

$$S(N,U) = k_B \ln \Omega(N,U). \tag{2.27}$$

When we expand $\Omega$ with the help of Stirling's formula, we find

$$S(N,U) = k_B\left\{\left(N + \frac{1}{2}\right)\ln N - \frac{1}{2}\ln(2\pi) + \frac{U}{2\mu B}\ln\left[\frac{N - \frac{U}{\mu B}}{N + \frac{U}{\mu B}}\right]\right.$$
$$\left. -\frac{1}{2}(N+1)\ln\left[\frac{1}{4}\left(N^2 - \left(\frac{U}{\mu B}\right)^2\right)\right]\right\}. \tag{2.28}$$

If we write $U = Nu$, where $u$ is the energy per spin, then we can simplify further (dropping the $\ln(2\pi)$ term which becomes negligible compared to the other terms in the $N \to \infty$ limit) to get the entropy as

$$S = Nk_B\left\{\ln 2 - \frac{1}{2}\ln\left[1 - \left(\frac{u}{\mu B}\right)^2\right] + \frac{u}{2\mu B}\ln\left[\frac{1 - \frac{u}{\mu B}}{1 + \frac{u}{\mu B}}\right]\right\}, \tag{2.29}$$

or in terms of $N$ and $U$:

$$S = Nk_B\left\{\ln 2 - \frac{1}{2}\ln\left[1 - \left(\frac{U}{\mu NB}\right)^2\right] + \frac{U}{2\mu NB}\ln\left[\frac{1 - \frac{U}{\mu NB}}{1 + \frac{U}{\mu NB}}\right]\right\}. \tag{2.30}$$

In the form above this is not that transparent – if we write the magnetization per spin as $m = M/N = -U/\mu NB$, then the entropy is

$$S = Nk_B\left[\ln 2 - \frac{1}{2}\ln\left(1 - m^2\right) - \frac{m}{2}\ln\left(\frac{1+m}{1-m}\right)\right]. \tag{2.31}$$

As a check of our calculation we can determine the value we expect for this expression in two limits where we can obtain the result by other means. In the limit $m \to 0$ the expression above gives $S = Nk_B\ln 2$, as we would expect, since all $2^N$ microstates are equally likely to occur, and when $m \to 1$, $S \to 0$, also as we would expect since there is only one microstate for which $M = N$.

Now, applying the relation

$$\frac{1}{T} = \left.\frac{\partial S}{\partial U}\right|_N \tag{2.32}$$

to Eq. (2.30), we get

$$
\begin{aligned}
\frac{1}{T} &= \left.\frac{dS}{dU}\right|_N \\
&= \frac{Nk_B}{2}\frac{2U}{(\mu NB)^2}\frac{1}{1 - \left(\frac{U}{\mu NB}\right)^2} + \frac{UNk_B}{2\mu NB}\left[-\frac{\frac{1}{\mu NB}}{1 - \frac{U}{\mu NB}} - \frac{\frac{1}{\mu NB}}{1 + \frac{U}{\mu NB}}\right] \\
&\quad + \frac{Nk_B}{2\mu NB}\ln\left(\frac{1 - \frac{U}{\mu NB}}{1 + \frac{U}{\mu NB}}\right) \\
&= \frac{k_B}{2\mu B}\ln\left(\frac{1 - \frac{U}{\mu NB}}{1 + \frac{U}{\mu NB}}\right),
\end{aligned}
\tag{2.33}
$$

which we can rewrite as

$$\frac{1 - \frac{U}{\mu NB}}{1 + \frac{U}{\mu NB}} = e^{\frac{2\mu B}{k_B T}}, \tag{2.34}$$

and when we solve for $U$ we get

$$
\begin{aligned}
U &= \mu NB\left(\frac{1 - e^{\frac{2\mu B}{k_B T}}}{1 + e^{\frac{2\mu B}{k_B T}}}\right) \\
&= -\mu NB \tanh\left(\frac{\mu B}{k_B T}\right).
\end{aligned}
\tag{2.35}
$$

We can use this to determine the magnetization per spin $m = -U/\mu NB$, which gives us

$$m = \tanh\left(\frac{\mu B}{k_B T}\right), \tag{2.36}$$

which is plotted in Fig. 2.4.

The curve for the magnetization per spin in Fig. 2.4 gives us considerable information about the equilibrium behaviour of the spin system as we vary the magnetic field $B$ and the temperature $T$. We can view $\mu B$ as a magnetic energy scale and $k_B T$ as a thermal energy scale, and the ratio of the two forms the dimensionless parameter $\mu B/k_B T$.[5] At low magnetic fields or high temperatures where $\mu B \ll k_B T$, there is essentially no net magnetization per spin, so the spins will be random in their orientation – there is not enough energy to be gained by aligning the spins with the field to overcome the thermal energy that allows spins to flip. At high magnetic fields or low temperatures $\mu B \gg k_B T$, the spins will be aligned with the applied field as there is not enough thermal energy to disorder them.

---

[5] Whenever possible, it is good practice to try to express physical quantities in terms of dimensionless parameters, as the limits on a dimensionless parameter often capture different physical behaviour.

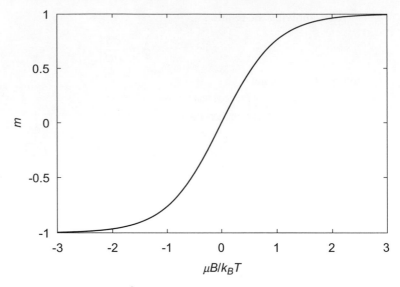

**Fig. 2.4**  Magnetization per spin as a function of $\frac{\mu B}{k_B T}$ for non-interacting spins.

We can also calculate the probability of finding a given spin pointing up, which will be

$$p_\uparrow = \frac{N_\uparrow}{N} = \frac{1}{2N}(N + M) = \frac{1}{2}(1 + m)$$

$$= \frac{1}{2}\left[1 + \tanh\left(\frac{\mu B}{k_B T}\right)\right]$$

$$= \frac{1}{2}\frac{e^{\frac{\mu B}{k_B T}} + e^{-\frac{\mu B}{k_B T}} + e^{\frac{\mu B}{k_B T}} - e^{-\frac{\mu B}{k_B T}}}{e^{\frac{\mu B}{k_B T}} + e^{-\frac{\mu B}{k_B T}}}$$

$$= \frac{e^{\frac{\mu B}{k_B T}}}{e^{\frac{\mu B}{k_B T}} + e^{-\frac{\mu B}{k_B T}}}, \tag{2.37}$$

and recalling that $p_\uparrow + p_\downarrow = 1$ we can also see that the probability that a spin is pointing down is

$$p_\downarrow = \frac{e^{-\frac{\mu B}{k_B T}}}{e^{\frac{\mu B}{k_B T}} + e^{-\frac{\mu B}{k_B T}}}. \tag{2.38}$$

The problem of the magnetization of non-interacting spins can also be solved in the canonical ensemble, see Problem 4.17.

## 2.5  Summary

In the microcanonical ensemble, the energy $U$, the particle number $N$ and the volume $V$ are fixed. Boltzmann's equation for the entropy

$$S = k_B \ln \Omega \qquad (2.39)$$

is an example of a *bridge equation* that connects a concept from thermodynamics (in this case the entropy $S$) to a statistical mechanical concept (here the multiplicity function $\Omega$). Once we have the thermodynamic quantity in terms of $\Omega$, we can use the regular machinery of thermodynamics to calculate physical properties of a system in equilibrium from parameters of the microscopic theory. A key feature of the microcanonical ensemble (from which the bridge equation follows) is that in equilibrium the probabilities of all the accessible microstates are equal, with $p_i = 1/\Omega$.

# Problems

**2.1**  In the card game bridge, each player has a hand with 13 cards drawn from a standard deck of 52 cards. What is the Shannon entropy associated with a bridge hand?

**2.2**  In a distant country, a politician suggests that car licence plates, which are six characters long: three random letters followed by three random digits between 0 and 9, should be replaced with a binary string of 0s and 1s. Use the Shannon entropy to determine how many characters will be needed for the new licence plates to encompass as many possibilities as the old licence plates.

**2.3**  In Section 2.1.2 we noted that the Schrödinger equation

$$i\hbar \frac{\partial \psi(\mathbf{x},t)}{\partial t} = -\frac{\hbar^2}{2m} \nabla^2 \psi(\mathbf{x},t) + V(x)\psi(\mathbf{x},t)$$

is time-reversal invariant. Demonstrate this explicitly by showing that if the equation has a solution $\psi(\mathbf{x},t)$, then there is a corresponding time-reversed solution $\psi^*(\mathbf{x},-t)$.

**2.4**  Consider two systems in thermal contact as in Section 2.1, one with energy $U_1$ and the other with energy $U_2 = U - U_1$.

(a) Show that the multiplicity function for the combined system is

$$\Omega(U,U_1) \simeq \Omega_1(\hat{U}_1)\Omega_2(\hat{U}_2) \exp\left[-\frac{(U_1 - \hat{U}_1)^2}{2\sigma_U^2}\right],$$

where

$$\sigma_U^2 = -\frac{k_B}{\left(\frac{\partial^2 S_1}{\partial U_1^2} + \frac{\partial^2 S_1}{\partial U_1^2}\right)},$$

and $S_1$ and $S_2$ are the entropies of systems 1 and 2, respectively.

(b) Show that

$$\frac{1}{k_B}\frac{\partial^2 S}{\partial U^2} = -\frac{1}{k_B T}\frac{1}{N c_v T},$$

where $c_v$ is the heat capacity per particle at constant volume. This equation implies that one can measure the heat capacity purely by observing equilibrium fluctuations at constant temperature.

(c) Assume that subsystem 1 has $N_1$ particles and subsystem 2 has $N_2$ particles. Write an expression for the relative width of energy fluctuations about equilibrium in terms of the heat capacities $c_V^{(1)}$ and $c_V^{(2)}$ of the two subsystems. If $N = N_1 + N_2 \to \infty$, comment on the behaviour of the fluctuations.

**2.5** A black hole is a region of space-time which has a sufficiently high matter density that the escape velocity exceeds the speed of light. A black hole of mass $M$ has a Schwarzschild radius

$$R_S = \frac{2GM}{c^2},$$

where $G$ is the gravitational constant and $c$ is the speed of light. Stephen Hawking showed that black holes emit blackbody radiation with a temperature

$$T_{\mathrm{BH}} = \frac{\hbar c^3}{8\pi G M k_B}.$$

(a) Given the result for the temperature of a black hole and recalling that the energy of the black hole is $E = Mc^2$, deduce that the entropy of a black hole is given by

$$S = \frac{k_B A}{4 l_P^2},$$

where $A$ is the area of the event horizon ($R_S$ is the radius of the event horizon) and $l_P = \left(G\hbar/c^3\right)^{\frac{1}{2}}$ is the Planck length.

(b) Viewing the entropy in terms of information, use the result from part (a) to calculate the number of bits that could be stored on the surface of a black hole of mass $M$. The black hole at the centre of the Milky Way galaxy, Sagittarius A*, has a mass of $\simeq 4 \times 10^6 M_\odot$, where $M_\odot \simeq 2 \times 10^{30}$ kg is the mass of the Sun. If it were possible to use this black hole as a hard drive, how many bits could be stored on its surface?

**2.6** Consider $N$ distinguishable atoms with two energy levels $\pm\epsilon$. Assume that there are very weak interactions between the atoms which allow equilibrium to be reached but that can be ignored for calculating equilibrium properties. If the total energy is $E$, what is the entropy of the system as a function of $E$? Use your answer to calculate the temperature of the system as a function of energy.

**2.7** A one-dimensional polymer may be modelled as being composed of $N$ monomers of length $a$ so that it has a maximum length of $L_{\max} = Na$. The entropy per link

$s = S/N$ is a sum of the configurational entropy $s_C$ per link and the entropy internal to a link, $s_I$:

$$s = s_I + s_C.$$

You are given that the heat capacity per link is $C_V = \alpha k_B$, where $\alpha$ is a constant.

(a) Show that the total entropy per link when the polymer has length $L$ is

$$s = k_B \left\{ \ln 2 - \frac{1}{2}(1 + x) \ln(1 + x) - \frac{1}{2}(1 - x) \ln(1 - x) + \alpha \ln T + C \right\},$$

where $x = L/L_{max}$ and $C$ is a constant.

(b) Calculate the change in temperature of the elastic band when the elastic band is stretched adiabatically from length 0 at initial temperature 300 K to length $L_{max}/2$.

**2.8** Recalling that Eq. (2.36) gives the magnetization per spin in units of $\mu$, show that at small magnetic fields, or high temperatures, the magnetization of a paramagnetic salt will satisfy Curie's law

$$M = \frac{CB}{T},$$

and find the value of the constant $C$.

**2.9** Consider $N$ identical simple harmonic oscillators with frequency $\omega$ for which the total energy is

$$E = \hbar\omega \left( n + \frac{N}{2} \right),$$

when $n$ excitations of energy $\hbar\omega$ are present. (The $\hbar\omega N/2$ term comes from the zero point energy of the oscillators.)

(a) Show that the number of ways of arranging the $n$ excitations between $N$ oscillators is

$$\Omega(N, n) = \frac{(N + n - 1)!}{N! \, (n - 1)!}.$$

(b) Using the result from part (a), find the entropy of the oscillators and hence use Stirling's approximation to show that the energy per oscillator at temperature $T$ is

$$\frac{U}{N} = \frac{\hbar\omega}{2} \coth\left( \frac{\hbar\omega}{2k_B T} \right).$$

We will obtain the same result using the canonical ensemble in Eq. (4.48).

**2.10** At high temperatures atoms in a crystal can be excited out of their regular positions in the crystal to interstitial positions that are not in the crystal lattice. Assume that there is an energy cost $\epsilon$ for an atom to move to an interstitial position, so that if $n$ interstitials are created, the energy is $U = n\epsilon$.

(a) If a crystal lattice has $N$ sites and $N$ possible interstitial positions, show that there are

$$\Omega(N,n) = \left[\frac{N!}{n!\,(N-n)!}\right]^2$$

ways for $n$ atoms to be excited to interstitial positions.

(b) Using the result from part (a), find the entropy of the system and use Stirling's approximation to find the energy $U$ as a function of temperature, and hence the heat capacity

$$C_V = \frac{dU}{dT}.$$

**2.11** Consider $N$ two-level systems that can be in one of two energy states, one with energy 0 and the other with energy $\epsilon$. Let the number of systems with energy 0 be $N_0$ and the number of systems with energy $\epsilon$ be $N_1$.

(a) Write down an expression relating $N$ to $N_0$ and $N_1$ and an expression relating the energy $U$ to $N_0$ and $N_1$. Hence calculate the entropy $S(U,N)$ and use Stirling's approximation to simplify it to

$$S = Nk_B \ln N - Nk_B \ln\left(N - \frac{U}{\epsilon}\right) + \frac{k_B U}{\epsilon} \ln\left(\frac{N - \frac{U}{\epsilon}}{\frac{U}{\epsilon}}\right).$$

(b) Find the temperature as a function of $U$. What is the value of the temperature when $U = 0$? When $U = N\epsilon/2$? When $U = 3N\epsilon/4$? Do any of these values strike you as surprising?

# Liouville's Theorem

There are a number of assumptions, some quite subtle, or not explicitly stated, that we have made up to this point that we have either not justified or not examined.

(1) A key assumption in the derivation of the Boltzmann entropy and a result that we showed from the Gibbs entropy in the microcanonical ensemble is that each allowed microstate (i.e. microstate within a certain narrow energy window) is equally probable. One thing we haven't worried about is whether in fact the system actually makes it into the microstates that entropy says it should in the most likely macrostate. It is not necessarily obvious that an allowed microstate is always accessible.

(2) Point (1) above is particularly important if a system is initialized far from equilibrium – how does a system relax to equilibrium? We might also ask whether a generic initial state will relax to equilibrium.

(3) We have discussed how to calculate physical quantities, at least in the microcanonical ensemble, but we have not worried about how these quantities are to be measured experimentally. In practice, when we make measurements of a system, we make measurements at many different times for the *same* system and average over these measurements, rather than measurements on many members of a statistical ensemble. How do we know that taking an ensemble average gives the same result as a temporal average? Can we show that the approach of using ensembles is valid for a system which is in only one particular microstate at a time?

These are important questions and we will attempt to provide some further insight into each of them, but it is certainly possible to address these questions in considerably more depth than our discussion here. We will focus on classical statistical mechanical systems for which we can frame the discussion in terms of phase space.

## 3.1 Phase Space and Hamiltonian Dynamics

Perhaps the most familiar example of phase space is for $N$ classical particles in three dimensions. In that case the phase space is $6N$-dimensional, corresponding to $6N$ variables: $3N$ position components $q_i$ and $3N$ momentum components $p_i$. If we were to know the values of all of these variables, then we could completely specify the microstate of the system $\{p_i, q_i\}$, which corresponds to a single point in phase space. In practice this is not usually possible, but conceptually it is useful to consider.

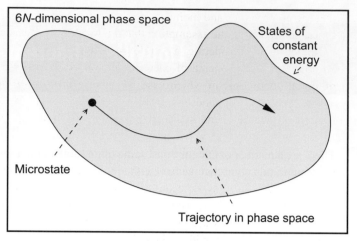

6N-dimensional phase space

States of
constant
energy

Microstate

Trajectory in phase space

**Fig. 3.1**     Schematic trajectory through phase space restricted to states with the same energy.

If we know the initial state $\{p_i, q_i\}$ and the Hamiltonian $H = T + V$ for classical particles, with $T$ the kinetic energy and $V$ the potential energy, then we can follow the subsequent time evolution via Hamilton's equations

$$\dot{q} = \frac{\partial H}{\partial p}; \qquad \dot{p} = -\frac{\partial H}{\partial q}, \tag{3.1}$$

and if we solve these then $p$ and $q$ are known for all later times. For example, for point particles with mass $m$ moving in a potential that depends only on the co-ordinates of the particles, the Hamiltonian takes the form

$$H\left(\{p_i, q_i\}\right) = \sum_{i=1}^{3N} \frac{p_i^2}{2m} + U\left(q_1, \ldots, q_{3N}\right),$$

and applying Hamilton's equations we get

$$\dot{q}_i = \frac{\partial H}{\partial p_i} = \frac{p_i}{m}, \tag{3.2}$$

$$\dot{p}_i = -\frac{\partial H}{\partial q_i} = -\frac{\partial U}{\partial q_i} = f_i(q_1, \ldots, q_{3N}), \tag{3.3}$$

where $f_i$ is a force (obtained from the spatial derivative of a potential energy). The equations of motion are just those of Newtonian dynamics, i.e.

$$\dot{p}_i = m\ddot{q}_i = f_i(q_1, \ldots, q_{3N}). \tag{3.4}$$

For a Hamiltonian which has no explicit time dependence, the Hamiltonian itself is a constant of the motion, which is equal to the initial energy $E$ of the system. Hence the trajectory in phase space from a particular initial state will be given by Hamiltonian dynamics and should only encompass microstates with energy $E$ (Fig. 3.1).

An important point is that trajectories in phase space never cross (Hamilton's equations have unique solutions). To see this, suppose that two distinct trajectories cross at time $t_0$. If we solve the equations of motion with initial conditions specified at $t = t_0$, this implies that the co-ordinates and momenta on the two trajectories will be equal at all subsequent times, contradicting the assumption that the two trajectories are distinct.

Rather than considering a single realization of the system, in which the initial co-ordinates and momenta of each particle are specified, we can consider an ensemble of initial conditions, in which case we have a probability density $\rho(\{p_i, q_i\})$ in phase space. We can understand this probability density as giving the probability $\rho \, d\mathbf{p} \, d\mathbf{q}$ that the system will be found in a microstate that lies in the $6N$-dimensional box of volume $d\mathbf{p} \, d\mathbf{q}$ centred on $[\mathbf{p}, \mathbf{q}]$, where $\mathbf{p}$ is a vector[1] consisting of all of the $p_i$ and similarly for $\mathbf{q}$. If the system is in a single microstate, the probability density consists of a $6N$-dimensional delta function. In practice we never know the initial conditions to infinite precision and a probability density allows the study of multiple closely related initial conditions, which is much more relevant to experiments than a delta function initial condition. The introduction of a probability density also allows for the possibility that the allowed microstates occupy a small energy window with width $\delta E$.

Now, probability cannot be created or destroyed, hence the probability density is *locally conserved*, it can only flow from one part of phase space to another. This implies that there is a phase-space probability current associated with the probability density that satisfies the continuity equation (one can view this as the regular continuity equation in three dimensions, $\frac{\partial \rho}{\partial t} + \boldsymbol{\nabla} \cdot (\rho \mathbf{v}) = 0$ generalized to $6N$ dimensions):

$$\frac{\partial \rho}{\partial t} + \sum_{i=1}^{3N} \left[ \frac{\partial (\rho \dot{q}_i)}{\partial q_i} + \frac{\partial (\rho \dot{p}_i)}{\partial p_i} \right] = 0. \tag{3.5}$$

Thus, when we apply the product rule to the terms in square brackets we get

$$\frac{\partial \rho}{\partial t} + \sum_{i=1}^{3N} \left[ \frac{\partial \rho}{\partial q_i} \dot{q}_i + \frac{\partial \rho}{\partial p_i} \dot{p}_i + \rho \left\{ \frac{\partial \dot{q}_i}{\partial q_i} + \frac{\partial \dot{p}_i}{\partial p_i} \right\} \right] = 0. \tag{3.6}$$

The terms in the curly brackets can be seen to vanish when we use Hamilton's equations:

$$\frac{\partial \dot{q}_i}{\partial q_i} + \frac{\partial \dot{p}_i}{\partial p_i} = \frac{\partial}{\partial q_i} \left( \frac{\partial H}{\partial p_i} \right) + \frac{\partial}{\partial p_i} \left( -\frac{\partial H}{\partial q_i} \right) = \frac{\partial^2 H}{\partial q_i \partial p_i} - \frac{\partial^2 H}{\partial q_i \partial p_i} = 0, \tag{3.7}$$

which leaves the equation

$$\frac{\partial \rho}{\partial t} + \sum_{i=1}^{3N} \left[ \frac{\partial \rho}{\partial q_i} \dot{q}_i + \frac{\partial \rho}{\partial p_i} \dot{p}_i \right] = 0, \tag{3.8}$$

and the left-hand side of the equation is $\frac{d\rho}{dt}$, hence

$$\frac{d\rho}{dt} = \frac{\partial \rho}{\partial t} + \sum_{i=1}^{3N} \left[ \frac{\partial \rho}{\partial q_i} \dot{q}_i + \frac{\partial \rho}{\partial p_i} \dot{p}_i \right] = 0, \tag{3.9}$$

---

[1] We will always use bold typeface to indicate vector quantities.

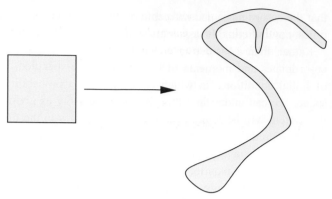

**Fig. 3.2**     Incompressible (volume-preserving) flow of probability density in phase space.

**Fig. 3.3**     An example of a pathological initial condition in phase space in which gas particles are lined up in a row moving with identical velocities.

which is **Liouville's theorem**. Now, $\frac{d\rho}{dt}$ is the total derivative of the probability density, i.e. the evolution of $\rho$ as seen by a particle moving with the flow, and the fact that $\frac{d\rho}{dt} = 0$ has the following important implications:

(1) The probability density is incompressible and hence volume in phase space is preserved, even if a volume element is considerably deformed in its evolution, as illustrated in Fig. 3.2.

(2) The system is generically not restricted to a small region of phase space (an attractor). In many dynamical systems, there is some initial transient behaviour and then the system explores only some small subset of states in phase space (the attractor), e.g. a forced oscillator will settle into periodic motion with a given frequency and not explore other regions of phase space. In statistical mechanics we have that the opposite is true – since Hamiltonian trajectories do not intersect, the system will eventually explore essentially all of the accessible states.

(3) In the microcanonical ensemble the energy is a constant of the motion, so Liouville's theorem implies that for essentially any initial state, the probability density will evolve to cover the entire hypersurface of constant energy uniformly so that there is non-

zero probability of finding the system in any region of phase space. Some initial states are pathological (e.g. gas molecules lined up in a row moving with identical velocity, Fig. 3.3), but these are isolated points in phase space and contain no volume.

## 3.2 Ergodic Hypothesis

The consequence of Liouville's theorem that in the limit of long times, systems uniformly sample all states with the same energy in phase space leads to the idea that a temporal average at fixed energy should sample accessible states with the same energy with uniform probability, which is exactly what we previously assumed in the microcanonical ensemble. This idea is summed up in the **ergodic hypothesis**, that the time spent by a trajectory in a particular region of phase space is proportional to the volume of that region of phase space. The consequence that is most important for us is that

*The temporal average of a thermodynamic variable is equivalent to an ensemble average of the same variable.*

The ergodic hypothesis allows a connection to be made between statistical mechanics (where we average over ensembles, e.g. by performing phase-space averages) and experiments (or simulations) where we average temporally. Mathematically we can express this connection for a quantity $f$ by asserting that the ensemble average

$$\langle f \rangle = \int_\Gamma f(x) d\mu(x), \tag{3.10}$$

where $\mu$ is the integration measure for phase space, is equal to the temporal average

$$\bar{f} = \lim_{T \to \infty} \frac{1}{T} \int_0^T dt\, f(x(t)), \tag{3.11}$$

where $x(t)$ is a representative trajectory through phase space (i.e. we don't start at some pathological point in phase space). In taking the temporal average, we allow the times that contribute to the average to be very large. This is to ensure that any transients associated with the initial conditions have time to decay, and is similar to the way we defined thermal equilibrium as a state that a system reaches at long times that is independent of initial conditions.

In general it is very difficult to prove ergodicity (although there are some special cases where it can be proven, e.g. hard spheres). However, Liouville's theorem motivates the idea that averages over representative regions of phase space should be equivalent to temporal averages along trajectories in phase space. While a rigorous proof of ergodicity is not generally possible, there are numerous systems for which it is possible to establish numerically that their behaviour is consistent with the ergodic hypothesis. This is achieved by comparing simulations of dynamics, e.g. using a molecular dynamics simulation to follow particle trajectories with a technique that samples phase space with equilibrium probabilities, e.g. Monte Carlo simulations. From a practical standpoint, the results from ensemble averages match well with those obtained from temporal averages and hence even if it is hard to justify statistical mechanics with full rigour, it is clear that it is a very useful approach to understanding the world.

### 3.2.1 Non-ergodic Systems

Not all systems satisfy the ergodic hypothesis – there are also systems which are non-ergodic, such as broken symmetry states, e.g. a ferromagnet, in which the system is only able to explore some subset of equal energy states, although it will explore this subset of states ergodically. Glassy systems also display broken ergodicity as they fall out of equilibrium when cooled (the relevant time scale for equilibration becomes so large as to be effectively infinite in the vicinity of the glass transition). It is an open question as to whether glassy phases break ergodicity if cooled infinitely slowly. A quantum example of a non-ergodic system is many-body localization, in which an isolated quantum system with strong interactions between degrees of freedom fails to reach equilibrium. Such systems retain memory of their initial state that can be observed in local measurements. There is a rigorous proof of the existence of this phase in one dimension for appropriate assumptions about the energy spectrum (Imbrie, 2016), but it is currently not known whether it can occur in higher-dimensional systems.

## 3.3 Summary

Thermalization in classical systems occurs via spreading of an initial probability distibution in phase space. That spreading is guaranteed to preserve phase-space volume for Hamiltonian systems by Liouville's theorem, and eventually covers phase space uniformly. This suggests the ergodic hypothesis that temporal averages of physical quantities are equivalent to ensemble averages.

## Problems

**3.1**  The position and momentum of a damped pendulum satisfy the equations (where you may assume $4k/m > \alpha^2$)
$$\dot{x} = \frac{p}{m},$$
$$\dot{p} = -\alpha p - k\sin(x).$$

(a) Assuming the amplitude of oscillations to be small, so that we can approximate $\sin(x)$ with $x$, find the limiting behaviour of $x$ and $p$ at long times – does the system explore all of phase space or is there an attractor for its dynamics?
(b) Why does Liouville's theorem not apply to the damped pendulum? Specifically consider $\frac{\partial \dot{p}}{\partial p}$ and $\frac{\partial \dot{q}}{\partial q}$ in the derivation of Liouville's theorem.

**3.2**  Consider a one-dimensional classical simple harmonic oscillator, which has Hamiltonian

$$H = \frac{p^2}{2m} + \frac{1}{2}\omega^2 x^2.$$

Obtain Hamilton's equations and then solve them to find $x$ and $p$ as a function of time. Peform a temporal average of the Hamiltonian to obtain the average energy.

In Chapter 4 we will see that the average energy for a classical one-dimensional simple harmonic oscillator is $k_B T$, which can be calculated from the equipartion theorem. Use this information to determine the amplitude of the motion of the oscillator when the temperature is $T$.

# 4    The Canonical Ensemble

In the microcanonical ensemble we considered a system that is thermally isolated from the rest of the world. Most thermodynamic systems that are of interest to us are not so well isolated – they are in contact with an environment (which we refer to as a heat bath or reservoir) that is much larger than the system of interest and can exchange energy with it (Fig. 4.1). We will not concern ourselves with the microscopic details of the bath, but assume that there are sufficiently many degrees of freedom in the bath that any amount of energy shifted to the system is negligible compared to the total energy in the bath and that there are sufficiently weak interactions between the bath and the system that they can be treated as distinct. We take the system of interest to be in thermodynamic equilibrium with the bath, which means that the temperature $T$ must be fixed, but the energy of the system, $E$, can fluctuate about its mean value $U = \langle E \rangle$. As we have seen earlier, the probability distribution for $U$ is such that the most probable values of $E$ are very close to $U$. We can treat the combination "bath + system of interest" as closed, with total energy $U_{\text{Total}}$, and assume that the mean energy in the system of interest $U \ll U_{\text{Total}}$.

## 4.1  Partition Function

To connect the statistical properties of the canonical ensemble to thermodynamics, we use a similar approach to the one that we used for the microcanonical ensemble. In the microcanonical ensemble we used the principle of maximum entropy and maximized the Gibbs entropy subject to the constraint on the sum of the probabilities $\{p_j\}$ of the microstates. We found that all microstates were equally probable, but did not assume this in our argument for the form of the Gibbs entropy. In the canonical ensemble there is an additional constraint that we need to consider, that the mean energy of the system has a well-defined value $U$ (we showed in Chapter 2 that the equilibrium energy $U$ is centred around a highly probable value). To take this into account in the same framework as before, we need to introduce a Lagrange multiplier $\beta$ in addition to $\lambda$ (introduced in Section 2.2) to capture that the average energy is a conserved quantity. We then maximize the Gibbs entropy (up to a factor of $k_B$, which does not affect our results for the $p_j$)

$$ S = -\sum_j p_j \ln p_j - \lambda \left( \sum_j p_j - 1 \right) - \beta \left( \sum_j \epsilon_j p_j - U \right), \tag{4.1} $$

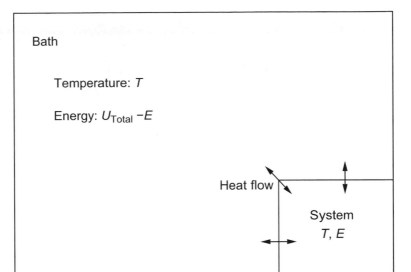

Bath

Temperature: $T$

Energy: $U_{Total} -E$

Heat flow

System
$T, E$

**Fig. 4.1** Schematic of the canonical ensemble: system plus heat bath.

where $\epsilon_j$ is the energy of microstate $j$. Similarly to the derivation of the bridge equation for the microcanonical ensemble, we require that the $\{p_j\}$ are chosen to maximize $S$ which leads to

$$- \left( \ln p_j + 1 \right) - \lambda - \beta \epsilon_j = 0, \tag{4.2}$$

which we can rearrange to give

$$p_j = e^{-(1+\lambda)-\beta\epsilon_j}. \tag{4.3}$$

Summing over all states on both sides of the equation and using the constraint $\sum_j p_j = 1$ gives

$$1 = e^{-(1+\lambda)} \sum_j e^{-\beta\epsilon_j}, \tag{4.4}$$

and hence we define

$$e^{(1+\lambda)} = \sum_j e^{-\beta\epsilon_j} = Z, \tag{4.5}$$

which fixes $\lambda$ and introduces the **partition function** as

$$Z = \sum_j e^{-\beta\epsilon_j}. \tag{4.6}$$

Equations (4.3) and (4.5), taken together, imply that

$$p_j = \frac{e^{-\beta\epsilon_j}}{Z}, \tag{4.7}$$

and so we can see that the role of the partition function is to ensure that $\sum_j p_j = 1$. We note that Eq. (4.7) reduces to the result we found in the microcanonical ensemble if all microstates have the same energy $\epsilon$, since in that case

$$Z = \Omega e^{-\beta\epsilon},\tag{4.8}$$

and then $p_j = 1/\Omega$. At this stage $\beta$ is a Lagrange multiplier, and we will comment on its physical significance shortly.

We notice that in the canonical ensemble, the probability of a microstate depends on its energy. The partition function ensures that these probabilities are properly normalized. We will see below that its usefulness comes from its relation to thermodynamic properties of the system.

## 4.2  Bridge Equation in the Canonical Ensemble

In the microcanonical ensemble we found a bridge equation connecting a thermodynamic variable, entropy, with a statistical concept, the multiplicity of states. We can find a bridge equation in the canonical ensemble by starting with the entropy and substituting the expression for the probability of microstate $j$, Eq. (4.7):

$$S = -k_B \sum_j p_j \ln p_j$$

$$= -k_B \sum_j \frac{e^{-\beta\epsilon_j}}{Z} \ln\left(\frac{e^{-\beta\epsilon_j}}{Z}\right)$$

$$= -k_B \sum_j \frac{e^{-\beta\epsilon_j}}{Z}\left(-\beta\epsilon_j\right) + k_B \sum_j \frac{e^{-\beta\epsilon_j}}{Z}\ln Z$$

$$= k_B\beta \sum_j \frac{\epsilon_j e^{-\beta\epsilon_j}}{Z} + k_B \ln Z \sum_j \frac{e^{-\beta\epsilon_j}}{Z}$$

$$= k_B\beta U + k_B \ln Z,\tag{4.9}$$

where we note

$$U = \sum_j \epsilon_j p_j = \frac{\sum_j \epsilon_j e^{-\beta\epsilon_j}}{Z},\tag{4.10}$$

which is a weighted average of the energies. We still have not determined $\beta$. We know that it can be varied so that we get the correct value of the mean energy for our system, but we have not given it a physical meaning.

To determine $\beta$, suppose there is a small reversible change in the average energy of the system $\delta U$, due to heat flowing into the system from the bath. This implies that there will be a change $\delta p_j$ in the probability $p_j$ of microstate $j$, so that we get a new set of probabilities $\{\tilde{p}_j\}$:

$$p_j \rightarrow \tilde{p}_j = p_j + \delta p_j,$$

and

$$\sum_j \tilde{p}_j = 1 \implies \sum_j (p_j + \delta p_j) = 1,$$

which, when combined with the condition $\sum_j p_j = 1$, implies that

$$\sum_j \delta p_j = 0, \tag{4.11}$$

which we will use several times below. Now, the change in energy leads to a change in the entropy of the system, $\delta S$, which we can calculate as follows:

$$
\begin{aligned}
\delta S &= S(\{\tilde{p}_j\}) - S(\{p_j\}) \\
&= -k_B \sum_j (p_j + \delta p_j) \ln \left( p_j + \delta p_j \right) + k_B \sum_j p_j \ln p_j \\
&= -k_B \sum_j p_j \ln \left( p_j + \delta p_j \right) - k_B \sum_j \delta p_j \ln \left( p_j + \delta p_j \right) + k_B \sum_j p_j \ln p_j, \tag{4.12}
\end{aligned}
$$

and

$$\ln \left( p_j + \delta p_j \right) = \ln p_j + \ln \left( 1 + \frac{\delta p_j}{p_j} \right), \tag{4.13}$$

hence the second term can be expanded to lowest order in $\delta p_j$ by using

$$\ln (1 + x) = x + \cdots , \tag{4.14}$$

to give

$$
\begin{aligned}
\delta S &= -k_B \sum_j \delta p_j \ln p_j - k_B \sum_j \delta p_j \\
&= k_B \sum_j \delta p_j \left( \beta \epsilon_j + \ln Z \right) \\
&= k_B \beta \sum_j \epsilon_j \delta p_j + k_B \ln Z \sum_j \delta p_j \\
&= k_B \beta \delta U. \tag{4.15}
\end{aligned}
$$

Hence in the limit of infinitesimal $\delta U$ we obtain

$$\frac{\delta S}{\delta U} \rightarrow \left. \frac{\partial S}{\partial U} \right|_{N,V} = k_B \beta = \frac{1}{T}, \tag{4.16}$$

where we recalled Eq. (2.11), which allows us to identify

$$\beta = \frac{1}{k_B T}. \tag{4.17}$$

We can now rewrite our equation involving $S$, $U$ and $Z$ as

$$TS = U + k_B T \ln Z,$$

or

$$U - TS = -k_B T \ln Z, \qquad (4.18)$$

and $U - TS = F$, the Helmholtz free energy, so our bridge equation is

$$F = -k_B T \ln Z. \qquad (4.19)$$

The equilibrium state may be found by minimizing the Helmholtz free energy (equivalent to maximizing the entropy with constraints) as we are used to in thermodynamics.

In our arguments above, we wrote expressions in terms of $U$. However, in practice, we usually do not know $U$ and measure the temperature $T$ from which the mean energy can be inferred. This is the way we will approach calculations.

### 4.2.1 Boltzmann Distribution

With the identification that $\beta = 1/k_B T$, an implication for the probabilities $p_j$ that we found above is that if we compare the probabilities of two states 1 and 2 with energies $\epsilon_1$ and $\epsilon_2$, respectively, then we get

$$\frac{p(\epsilon_1)}{p(\epsilon_2)} = \frac{\exp\left\{-\frac{\epsilon_1}{k_B T}\right\}}{\exp\left\{-\frac{\epsilon_2}{k_B T}\right\}} = \exp\left\{-\frac{(\epsilon_1 - \epsilon_2)}{k_B T}\right\}, \qquad (4.20)$$

which is known as the Boltzmann distribution, with the individual exponentials

$$\exp\left\{-\frac{\epsilon}{k_B T}\right\},$$

known as Boltzmann factors. The probabilities determined in Eq. (4.7) are normalized Boltzmann factors, and exploiting the relationship between the partition function $Z$ and the Helmholtz free energy we may write them as

$$p(\epsilon_j) = \exp\left\{\frac{F - \epsilon_j}{k_B T}\right\}. \qquad (4.21)$$

### 4.2.2 Derivatives of the Partition Function

Once we know the Helmholtz free energy, there are many thermodynamic quantities we can determine through use of partial derivatives, making use of our knowledge of thermodynamics. These can also be related directly to derivatives of the partition function $Z$, as we see below.

The mean energy is

$$U = \langle E \rangle = \sum_j \epsilon_j p_j = \frac{\sum_j \epsilon_j e^{-\beta \epsilon_j}}{Z}, \qquad (4.22)$$

which can be written in the form

$$U = -\frac{1}{Z}\frac{\partial Z}{\partial \beta} = -\frac{\partial \ln Z}{\partial \beta} = k_B T^2 \frac{\partial \ln Z}{\partial T} = k_B T^2 \frac{1}{Z}\frac{\partial Z}{\partial T} , \tag{4.23}$$

where we noted

$$\frac{\partial}{\partial \beta} = \frac{\partial T}{\partial \beta}\frac{\partial}{\partial T} = \frac{1}{\frac{\partial \beta}{\partial T}}\frac{\partial}{\partial T} = -k_B T^2 \frac{\partial}{\partial T} . \tag{4.24}$$

We can evaluate the heat capacity in a similar manner:

$$C_V = \frac{\partial U}{\partial T}\bigg|_V$$

$$= \frac{\partial}{\partial T}\left[\frac{1}{Z}\sum_j \epsilon_j e^{-\beta \epsilon_j}\right]$$

$$= \frac{1}{Z}\sum_j \frac{\epsilon_j^2}{k_B T^2} e^{-\beta \epsilon_j} - \frac{1}{Z^2}\frac{\partial Z}{\partial T}\sum_j \epsilon_j e^{-\beta \epsilon_j}$$

$$= \frac{1}{k_B T^2}\left\langle E^2\right\rangle - \frac{\partial \ln Z}{\partial T}\frac{1}{Z}\sum_j \epsilon_j e^{-\beta \epsilon_j}$$

$$= \frac{1}{k_B T^2}\left(\left\langle E^2\right\rangle - \langle E\rangle^2\right), \tag{4.25}$$

where we used Eq. (4.23) to relate the average energy $U$ to a derivative of the partition function.

Equation (4.25) is an example of a relation between a macroscopic susceptibility ($C_V$, the linear response of the energy to a temperature change) and equilibrium microscopic fluctuations of the energy. This is actually very general behaviour in equilibrium and is a particular example of the *fluctuation–dissipation theorem*. The significance of the result in this particular example is that it means one can determine the heat capacity of a system without ever changing its temperature, simply by observing equilibrium fluctuations in the energy.

## Entropy and Pressure

We already have an expression for the entropy in terms of the multiplicity function $\Omega$, but if we want to calculate the entropy in the canonical ensemble, then it is useful to know the following expression:

$$S = -\frac{\partial F}{\partial T}\bigg|_{N,V} , \tag{4.26}$$

which we can check easily using the bridge equation:

$$-\frac{\partial F}{\partial T} = \frac{\partial}{\partial T}\left(k_B T \ln Z\right)$$

$$= k_B \ln Z + \frac{k_B T}{Z}\frac{\partial Z}{\partial T}$$

$$= -\frac{F}{T} + \frac{U}{T}$$

$$= S. \tag{4.27}$$

Another quantity of interest that has not arisen in the examples we have considered so far is that of pressure (obviously of interest for discussing gases). This can be obtained from partial derivatives of either the internal energy (see Eq. (B.18)):

$$P = -\left.\frac{\partial U}{\partial V}\right|_{N,S}, \tag{4.28}$$

or the Helmholtz free energy (see Eq. (B.32)):

$$P = -\left.\frac{\partial F}{\partial V}\right|_{N,T}. \tag{4.29}$$

We can substitute Eq. (4.19) into Eq. (4.29) and then

$$
\begin{aligned}
P &= \frac{\partial}{\partial V} k_B T \ln Z \\
&= k_B T \frac{1}{Z} \frac{\partial Z}{\partial V} \\
&= k_B T \frac{1}{Z} \sum_n \frac{\partial}{\partial V} e^{-\beta \epsilon_n} \\
&= \frac{1}{Z} \sum_n \left(-\frac{\partial \epsilon_n}{\partial V}\right) e^{-\beta \epsilon_n},
\end{aligned} \tag{4.30}
$$

so from a statistical mechanical point of view, pressure reflects the dependence of the allowed energies on system size.

### 4.2.3 Equivalence of the Canonical and Microcanonical Ensembles

We have introduced the statistical ensembles as a way to relate microscopic degrees of freedom to macroscopic thermodynamic degrees of freedom. As such, we would hope that the relations we get for thermodynamic quantities should be independent of the method we use to obtain them. In contrast, it is not obvious that the microcanonical and canonical ensembles should give the same results, since for the former, the energy of the system is fixed, whereas for the latter, the energy of the system can fluctuate. However, by rearranging Eq. (4.25), we can look at relative fluctuations of the energy about the mean in the canonical ensemble:

$$\frac{\Delta E}{\langle E \rangle} = \frac{\sqrt{\langle E^2 \rangle - \langle E \rangle^2}}{\langle E \rangle} = \frac{\sqrt{k_B T^2 C_V}}{\langle E \rangle} \sim \frac{1}{\sqrt{N}}, \tag{4.31}$$

and we see that since $C_V$ and $\langle E \rangle$ are both extensive quantities and scale with $N$, the relative size of fluctuations tends to zero in the $N \to \infty$ limit, so the probability of finding the system with an energy other than $\langle E \rangle$ vanishes as $N \to \infty$. Hence even though the energy is allowed to fluctuate, in practice, the energy in the canonical ensemble for sufficiently large systems takes a well-defined value and takes the same value as that obtained in the microcanonical ensemble. Thus, averages of thermodynamic quantities in the canonical ensemble are dominated by the same set of states used for averages in the microcanonical ensemble, and will give the same results in the $N \to \infty$ limit.

# 4.3 Connections to Thermodynamics

We have already seen that thermodynamics and statistical mechanics are intimately related, but it is worth exploring the connection between the two in more depth. The key physics that comes from statistical mechanics is that it provides a microscopic basis for thermodynamics – using knowledge of the microscopic degrees of freedom, the microstates, one can calculate statistical properties of the system that can then be related to thermodynamic quantities. We have seen this already in the bridge equations that relate statistical and thermodynamic quantities for the microcanonical and canonical ensembles. Thermodynamics is a theory of macrostates without consideration of microscopic degrees of freedom, and statistical mechanics connects the microstates with macrostates. In particular, thermodynamics is a theory for $N \to \infty$, a limit often referred to as the *thermodynamic limit*. In statistical mechanics we find that as $N \to \infty$ relative fluctuations in thermodynamic quantities $\sim 1/\sqrt{N}$, e.g. in Eq. (4.31), so thermodynamic variables have well-defined limits when $N$ becomes large. However, statistical mechanics also allows one to focus on systems with finite $N$, and its methods are applicable to problems in information theory, economics and complexity theory, for instance, where the emergent behaviour in the limit of large $N$ does not have an obvious connection to thermodynamics.

Whilst statistical mechanics provides a microscopic justification for thermodynamics, thermodynamics is itself a logically self-contained theory, based on the three laws, supplemented by the zeroth law. We briefly review each of the laws in light of what we have learned thus far.

(0) **Zeroth law**: *If two systems are in equilibrium with a third, they are in equilibrium with each other.*

We saw that when two systems are in thermal contact in equilibrium, requiring the energy to be distributed in the most probable way implies that $\frac{\partial S}{\partial U}$ takes a constant value in both systems, which we identify as $1/T$. Thus, the two systems in equilibrium have the same value of temperature. We can imagine bringing the two systems in equilibrium in contact with a third, and the condition for equilibrium would thus be that all three systems have the same temperature, consistent with the statement of the zeroth law.

(1) **First law**: *Conservation of energy.*

We assume that conservation of energy holds at the microscopic level, which then implies that it must also hold at the macroscopic level. We can argue for this by contradiction – if it didn't hold at the microscopic level, we would have no reason to believe it should hold at the macroscopic level.

(2) **Second law**: *Entropy always increases.*
When we investigated two systems in thermal contact, we saw that after we removed an internal constraint, the entropy increased because $\Omega_f > \Omega_i$.

(3) **Third law**: *The entropy per particle goes to zero at absolute zero.*
This follows from the statistical definition of entropy

$$S(T = 0) = k_B \ln \Omega_0, \qquad (4.32)$$

where $\Omega_0$ is the ground-state degeneracy. In most quantum systems, the ground-state degeneracy is finite, in which case the entropy per particle $S/N$ goes to zero, affirming the third law. There are systems with macroscopically many low-energy states, such as glasses (although glasses are not in equilibrium) and certain spin systems on special lattices (e.g. an antiferromagnet[1] on a kagome lattice), which have macroscopically many classical ground states (however, such degeneracy is very susceptible to being broken by small perturbations).

## 4.4 Examples

Having introduced the formalism of the canonical ensemble, we now demonstrate its application in a variety of examples. We will consider one-particle examples for which we will obtain the one-particle partition function $Z_1$. Given our previous discussion that the thermodynamic limit applies as $N \to \infty$, it might seem odd to consider $N = 1$. However, we will see in Section 4.5.2 how to relate these one-particle results to partition functions for systems with $N \gg 1$ and how to obtain results that apply in the thermodynamic limit. The examples we consider, a two-level system, the simple harmonic oscillator, rigid rotors and a particle in a box, will turn out to allow us to give a statistical mechanical description of a gas of polyatomic molecules.

### 4.4.1 Two-Level System

Consider a particle with two energy levels at 0 and $\epsilon$, as illustrated in Fig. 4.2. This is about as simple a statistical mechanical system as one can have, and is a simple model for defects in solids. The partition function is straightforward to calculate:

$$Z = \sum_{n=0}^{1} e^{-\beta \epsilon_n} = 1 + e^{-\beta \epsilon}. \qquad (4.33)$$

It is also straightforward to calculate the average energy

$$U = \frac{0 + \epsilon e^{-\beta \epsilon}}{Z} = \frac{\epsilon e^{-\beta \epsilon}}{1 + e^{-\beta \epsilon}}. \qquad (4.34)$$

---

[1] We will discuss antiferromagnets in Chapter 10.

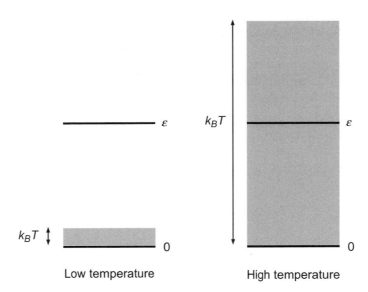

Energy-level diagram for a two-level system. The shaded areas for low and high temperatures indicate energies accessible via thermal fluctuations of order $k_BT$.

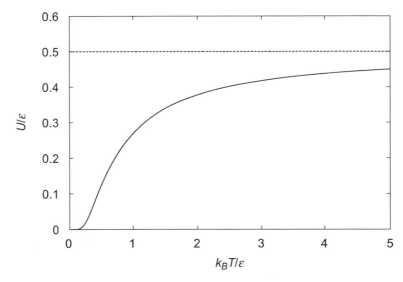

Dimensionless energy as a function of dimensionless temperature for a two-level system.

We plot Eq. (4.34) in Fig. 4.3. We can see that at high temperatures ($\beta \to 0$), the energy saturates to $U = \epsilon/2$. We can understand this easily by reference to Fig. 4.2. For high temperatures, the thermal energy available is $k_BT \gg \epsilon$, so the difference in the energy of the levels is not an impediment to their occupation. In this limit we expect the probability of the system being in either level to be $\sim 1/2$, and hence the average energy will be $\epsilon/2$. At low temperatures, the probability of excitation to the higher energy level will be exponentially suppressed, and so the energy will grow very slowly with temperature until $k_B T$ is an appreciable fraction of $\epsilon$ and the upper level has non-negligible probability of being occupied.

We can also calculate the heat capacity at constant volume

$$C_V = \left. \frac{\partial U}{\partial T} \right|_V, \tag{4.35}$$

which is

$$\begin{aligned}
C_V &= \frac{\partial}{\partial T} \left[ \frac{\epsilon e^{-\beta \epsilon}}{1 + e^{-\beta \epsilon}} \right] \\
&= \epsilon \frac{\partial}{\partial T} \left[ \frac{1}{1 + e^{\beta \epsilon}} \right] \\
&= \epsilon \left( \frac{\epsilon}{k_B T^2} \right) \frac{e^{\beta \epsilon}}{(1 + e^{\beta \epsilon})^2} \\
&= k_B \left( \frac{\epsilon}{k_B T} \right)^2 \frac{e^{-\beta \epsilon}}{(1 + e^{-\beta \epsilon})^2},
\end{aligned} \tag{4.36}$$

which we can write in terms of a function of a single dimensionless parameter:

$$\frac{C_V}{k_B} = f \left( \frac{\epsilon}{k_B T} \right), \tag{4.37}$$

where

$$f(x) = \frac{x^2 e^{-x}}{(1 + e^{-x})^2}. \tag{4.38}$$

From Eqs (4.37) and (4.38) we can determine several asymptotic limits. At low temperatures, $k_B T \ll \epsilon$, the heat capacity is exponentially small:

$$C_V \sim k_B \left( \frac{\epsilon}{k_B T} \right)^2 e^{-\beta \epsilon}, \tag{4.39}$$

because the only way to change the energy of the system is to increase the occupation probability of the higher level, which is exponentially small at low temperatures due to the presence of an energy gap to the higher level. At high temperatures, $k_B T \gg \epsilon$,

$$C_V \sim k_B \left( \frac{\epsilon}{2 k_B T} \right)^2, \tag{4.40}$$

which becomes very small since in the high-temperature limit both states become essentially equally probable, so $U$ changes very slowly with $T$. The occupation probability of the higher-energy state must always be less than that of the lower-energy state, hence $C_V$ must be very small. We plot the full expression for the heat capacity in Fig. 4.4. In the region between the low- and high-temperature limits, there is a peak which is an example of a Schottky anomaly – a peak in the heat capacity at low temperatures – which occurs in the temperature range between 0 and $\epsilon$, when the probability for excitation to the upper level grows rapidly with temperature.

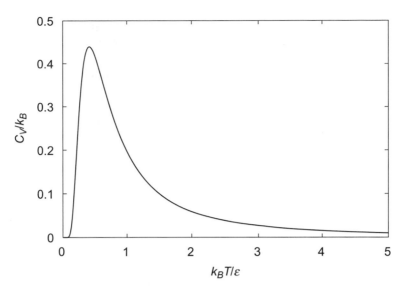

Fig. 4.4 Heat capacity as a function of dimensionless temperature for a two-level system. The peak for $k_BT/\epsilon \sim 0.4$ is an example of a Schottky anomaly.

### 4.4.2 Quantum Simple Harmonic Oscillator

The energy levels for the quantum simple harmonic oscillator (SHO) are given by the eigenvalues of the simple harmonic oscillator Hamiltonian:

$$H = \frac{p^2}{2m} + \frac{1}{2}kx^2, \tag{4.41}$$

and take the values

$$\epsilon_j = \hbar\omega\left(j + \frac{1}{2}\right) \tag{4.42}$$

for integers $j \geq 0$ with $\omega = \sqrt{k/m}$. Hence the corresponding probability for the state labelled $j$ is

$$p_j = \frac{e^{-\beta\hbar\omega\left(j+\frac{1}{2}\right)}}{Z}, \tag{4.43}$$

and the partition function can be calculated by noting that it has the form of an infinite geometric series:

$$Z = \sum_{j=0}^{\infty} e^{-\beta\epsilon_j}$$

$$= \sum_{j=0}^{\infty} e^{-\beta\hbar\omega\left(j+\frac{1}{2}\right)}$$

$$= e^{-\frac{\beta\hbar\omega}{2}} \sum_{j=0}^{\infty} e^{-j\beta\hbar\omega}$$

$$= \frac{e^{-\frac{\beta\hbar\omega}{2}}}{1 - e^{-\beta\hbar\omega}}$$

$$= \frac{1}{e^{\frac{\beta\hbar\omega}{2}} - e^{-\frac{\beta\hbar\omega}{2}}}$$

$$= \frac{1}{2 \sinh\left(\frac{\hbar\omega}{2k_BT}\right)}, \tag{4.44}$$

where we recalled that

$$\sum_{n=0}^{\infty} x^n = \frac{1}{1-x}, \tag{4.45}$$

for $|x| < 1$. Once we have the partition function we can calculate all other thermodynamic quantities of interest. If we are interested in the average energy $U$ then we can obtain it by taking derivatives of the partition function, as we showed in Eq. (4.23), or by direct calculation. We will demonstrate a direct calculation here, but note that it is simpler to use $U = -\frac{1}{Z}\frac{\partial Z}{\partial \beta}$. We start from the definition of $U$ and then we can calculate that

$$U = \sum_{j=0}^{\infty} \epsilon_j p_j = \hbar\omega \sum_{j=0}^{\infty} \frac{\left(j + \frac{1}{2}\right) e^{-\beta\hbar\omega\left(j+\frac{1}{2}\right)}}{Z}$$

$$= \hbar\omega \left[\frac{e^{-\frac{\beta\hbar\omega}{2}}}{Z} \sum_{j=0}^{\infty} j e^{-j\beta\hbar\omega} + \frac{1}{2}\right], \tag{4.46}$$

and we can use the trick that

$$\sum_{j=0}^{\infty} j e^{-j\beta\hbar\omega} = -\left[\frac{\partial}{\partial x} \sum_{j=0}^{\infty} e^{-jx}\right]_{x=\beta\hbar\omega}$$

$$= -\frac{\partial}{\partial x}\left[\frac{1}{1 - e^{-x}}\right]_{x=\beta\hbar\omega}$$

$$= \left[\frac{e^{-x}}{(1 - e^{-x})^2}\right]_{x=\beta\hbar\omega}. \tag{4.47}$$

Hence, making use of the expression we found for $Z$ in deriving Eq. (4.44), we obtain

$$U = \frac{\hbar\omega}{2} + \frac{\hbar\omega e^{-\beta\hbar\omega}}{1 - e^{-\beta\hbar\omega}}$$

$$= \frac{\hbar\omega}{2} + \frac{\hbar\omega}{e^{\beta\hbar\omega} - 1}$$

$$= \frac{\hbar\omega}{2}\left[\frac{e^{\beta\hbar\omega} + 1}{e^{\beta\hbar\omega} - 1}\right]$$

$$= \frac{\hbar\omega}{2} \coth\left(\frac{\hbar\omega}{2k_BT}\right). \tag{4.48}$$

If we view the energy of the quantum SHO as determined by exciting quanta of an oscillator, so that an energy $\epsilon_n$ corresponds to exciting $n$ quanta with energy $\hbar\omega$, then the mean number of excitations in the mode will thus be given by

$$\langle n_{\hbar\omega} \rangle = \frac{U - \frac{1}{2}\hbar\omega}{\hbar\omega} = \frac{1}{e^{\beta\hbar\omega} - 1}. \tag{4.49}$$

As we shall see in Chapter 9, this formula also arises in the context of calculating Planck's formula for the energy density of blackbody radiation, and can be related to the mean occupation of photon modes in a cavity.

### 4.4.3 Classical Partition Function and Classical Harmonic Oscillator

The SHO gives an example where we can compare the result obtained from the discrete spectrum of a quantum problem to that obtained from the continuous spectrum of the equivalent classical problem. In writing down the partition function for the classical harmonic oscillator, we start from the energies

$$E(p, x) = \frac{p^2}{2m} + \frac{1}{2}kx^2. \tag{4.50}$$

The microstates of the one-dimensional SHO are given by points in two-dimensional classical phase space, and so we replace a discrete sum by an integration over phase space and define the classical partition function by an integral over phase space:

$$Z_{\text{classical}} = \int_{-\infty}^{\infty} \int_{-\infty}^{\infty} \frac{dp\,dx}{2\pi\hbar} e^{-\beta E(p, x)}. \tag{4.51}$$

The integration over phase space is the classical equivalent of a sum over discrete energy levels in a quantum system. It might seem strange that we have introduced the factor $2\pi\hbar = h$ in the integral.[2] There are, however, two important reasons for this: (i) it makes the integral dimensionless and (ii) for consistency with the quantum problem (we will show this below). The factor $1/h = 1/(2\pi\hbar)$ has no consequence for calculating averages of physical quantities, since when we calculate averages there will be a factor in the numerator as well as the denominator ($Z$) and the factors of $h$ will cancel.

When we insert the expression for the energy of the SHO, Eq. (4.50), into the classical partition function Eq. (4.51), we get

$$Z_{\text{classical}} = \int_{-\infty}^{\infty} \int_{-\infty}^{\infty} \frac{dp\,dx}{2\pi\hbar} e^{-\frac{\beta p^2}{2m}} e^{-\frac{\beta k x^2}{2}}. \tag{4.52}$$

We can perform the Gaussian integrals immediately using Eq. (A.2) to get

$$Z_{\text{classical}} = \frac{1}{2\pi\hbar} \sqrt{2\pi k_B T m} \sqrt{\frac{2\pi k_B T}{k}} = \frac{k_B T}{\hbar\omega}. \tag{4.53}$$

To compare with the quantum problem, we take the classical limit of $Z_{\text{quantum}}$, which corresponds to the case where the available thermal energy is much larger than the separation between adjacent oscillator levels, $k_B T \gg \hbar\omega$, i.e. $\beta\hbar\omega \ll 1$, in which case

---

[2] We include a factor of $2\pi\hbar$ for each pair $dp\,dx$, so for example in three dimensions where we integrate over $d^3\mathbf{p}\,d^3\mathbf{x}$, there is a factor of $(2\pi\hbar)^3$.

$$Z_{\text{quantum}} = \frac{e^{-\frac{\beta\hbar\omega}{2}}}{1 - e^{-\beta\hbar\omega}}$$

$$\longrightarrow \frac{1}{1 - (1 - \beta\hbar\omega + \cdots)}$$

$$= \frac{k_B T}{\hbar\omega}. \tag{4.54}$$

We see that the classical limit of $Z_{\text{quantum}}$ is indeed $Z_{\text{classical}}$, validating the normalization we chose for the classical partition function.

### 4.4.4  Rigid Rotor

Molecules can have rotational degrees of freedom, and in general the energy eigenvalues associated with rotations depend on the distribution of mass in the molecule. The Hamiltonian for a rigid rotor is

$$H_{\text{rigid rotor}} = \frac{J_a^2}{2I_a} + \frac{J_b^2}{2I_b} + \frac{J_c^2}{2I_c}, \tag{4.55}$$

which depends on the angular momentum components $J_a = \mathbf{J} \cdot \hat{\mathbf{a}}$, $J_b = \mathbf{J} \cdot \hat{\mathbf{b}}$ and $J_c = \mathbf{J} \cdot \hat{\mathbf{c}}$, where $\hat{\mathbf{a}}$, $\hat{\mathbf{b}}$ and $\hat{\mathbf{c}}$ are the principal axes of the molecule, and the principal moments of inertia are $I_a$, $I_b$ and $I_c$. In general, when $I_a \neq I_b \neq I_c$, this problem does not have an analytic solution, but analytic expressions for the energies can be found for high-symmetry situations. Three of these higher-symmetry situations are the spherical rotor ($I_a = I_b = I_c$), the symmetric rotor ($I_a = I_b \neq I_c$) and the linear rotor ($I_a = I_b$, $I_c \to 0$). We discuss each of these examples in turn below.

When we calculate the partition function and sum over states, we need to count all of the states that have the same energy. In such a situation we can write the partition function in the form

$$Z = \sum_j g_j e^{-\beta\epsilon_j}, \tag{4.56}$$

where $g_j$ is the degeneracy of states with energy $\epsilon_j$. The degeneracy depends on the symmetry of the rotor. While superficially the form of the partition function as written in Eq. (4.56) appears to be different from the form we introduced earlier in Eq. (4.6), their physical content is the same. In Eq. (4.6) we introduced the partition function as a sum over states. However, we can group states with the same energy together, in which case the partition function may be rewritten as Eq. (4.56). It is important to be aware that both forms of the partition function are widely used – we will try to be clear as to which form we are using at any particular time.

### Spherical Rotor

The spherical rotor corresponds to a situation in which all three moments of inertia are equal: $I_a = I_b = I_c = I$. Examples of molecules that can be described as spherical rotors include methane ($CH_4$), uranium hexafluoride ($UF_6$) and buckyballs ($C_{60}$). The energy eigenvalues for a spherical rotor are

$$\epsilon_j^{\text{spherical rotor}} = \frac{\hbar^2}{2I} j(j+1), \tag{4.57}$$

where $j = 0, 1, 2, \ldots$ is a non-negative integer. The degeneracy of these levels is $g_j = (2j+1)^2$.

## Symmetric Rotor

For the symmetric rotor, two moments of inertia are the same, and one differs (e.g. $I_a = I_b \neq I_c$). Examples of molecules that can be described as symmetric rotors include ammonia ($NH_3$) and benzene ($C_6H_6$). In this case the energy eigenvalues are

$$\epsilon_{j,k}^{\text{symmetric rotor}} = \frac{\hbar^2}{2I_a} j(j+1) + \frac{\hbar^2}{2} \left( \frac{1}{I_c} - \frac{1}{I_a} \right) k^2, \tag{4.58}$$

where the solution is now labelled by two integers, $j$ and $k$. The allowed values of $j$ run over non-negative integers as for the spherical rotor, and in addition, the allowed values of $k$ from $-j, -j+1, \ldots, 0, \ldots, j-1, j$. Thus for each energy eigenvalue the degeneracy is $g_j = 2(2j+1)$ if $k \neq 0$ and $2j+1$ if $k = 0$.

## Linear Rotor

We can view the linear rotor as the $I_c \to 0$ limit of the symmetric rotor, corresponding e.g. to a diatomic molecule, or a linear molecule such as $CO_2$. In this case, the only allowed value of $k$ is $k = 0$, and with $I_a = I_b = I$ the energy eigenvalues are

$$\epsilon_j^{\text{linear rotor}} = \frac{\hbar^2}{2I} j(j+1), \tag{4.59}$$

with a degeneracy of $g_j = 2j+1$. In what follows, we will only consider linear rotors when discussing molecular rotations.

For the linear rotor the partition function is

$$Z = \sum_{j=0}^{\infty} (2j+1) \exp\left\{ -\frac{\hbar^2 j(j+1)}{2Ik_BT} \right\}. \tag{4.60}$$

We can rewrite this as

$$Z = \sum_{j=0}^{\infty} (2j+1) \exp\left\{ -\frac{\theta_r j(j+1)}{T} \right\}, \tag{4.61}$$

where

$$\theta_r = \frac{\hbar^2}{2Ik_B} \tag{4.62}$$

is a characteristic temperature for rotations. We cannot evaluate this partition function precisely, but we can investigate the behaviour in the low- and high-temperature limits.

At low temperatures, $T \ll \theta_r$, only the first few terms in the partition function are non-negligible, so we may approximate

$$Z \simeq 1 + 3 \exp \left\{ -\frac{2\theta_r}{T} \right\}, \tag{4.63}$$

and at high temperatures, $T \gg \theta_r$, the exponential decays very slowly, so the summand varies slowly as a function of $j$ and many terms in the sum contribute; we may approximate the sum by an integral, in which case we obtain

$$\begin{aligned} Z &\simeq \int_0^\infty dj\, (2j+1) \exp \left\{ -\frac{\theta_r j(j+1)}{T} \right\} \\ &= \left[ -\frac{T}{\theta_r} \exp \left\{ -\frac{\theta_r j(j+1)}{T} \right\} \right]_0^\infty \\ &= \frac{T}{\theta_r}. \end{aligned} \tag{4.64}$$

## 4.4.5  Particle in a Box

The energy levels for a particle in a one-dimensional box of length $L$ are

$$E_n = \frac{\hbar^2}{2m} \frac{\pi^2}{L^2} n^2, \tag{4.65}$$

where $n$ is a positive integer. Similarly, for a three-dimensional cube of side length $L$, we can solve the Schrödinger equation using separation of variables to find that the allowed energy levels are

$$E_{n_x,n_y,n_z} = \frac{\hbar^2 \pi^2}{2mL^2} \left( n_x^2 + n_y^2 + n_z^2 \right), \tag{4.66}$$

with $n_x$, $n_y$, $n_z$ all positive integers. We can evaluate the partition function as

$$\begin{aligned} Z &= \sum_{n_x,n_y,n_z=1}^\infty \exp \left\{ -\frac{\hbar^2 \pi^2}{2mk_B T L^2} \left( n_x^2 + n_y^2 + n_z^2 \right) \right\} \\ &= \sum_{n_x,n_y,n_z=1}^\infty \exp \left\{ -\frac{\theta_t}{T} \left( n_x^2 + n_y^2 + n_z^2 \right) \right\} \\ &= \left( \sum_{n_x=1}^\infty \exp \left\{ -\frac{\theta_t n_x^2}{T} \right\} \right)^3, \end{aligned} \tag{4.67}$$

where we introduced the characteristic temperature for translational motion:

$$\theta_t = \frac{\hbar^2 \pi^2}{2mk_B L^2}, \tag{4.68}$$

and we noted that the sums over $n_x$, $n_y$ and $n_z$ are identical. In the limit of high temperatures, i.e. $T \gg \theta_t$ (this is actually fairly easily achieved, for example for $H_2$ gas in a container of size $L = 1\,\mu$m, $\theta_t = 1.2\,\mu$K and $\theta_t$ decreases with increasing box size), we may convert the sums to integrals as we did when considering rotations and hence

$$Z = \left( \int_0^\infty dn \, \exp\left\{ -\frac{\theta_t}{T} n^2 \right\} \right)^3$$

$$= \left( \frac{1}{2} \sqrt{\frac{\pi T}{\theta_t}} \right)^3$$

$$= \left( \frac{\pi T}{4\theta_t} \right)^{\frac{3}{2}}$$

$$= \left( \frac{m k_B T}{2\pi \hbar^2} \right)^{\frac{3}{2}} V$$

$$= n_Q V, \tag{4.69}$$

where we wrote $L^3 = V$ and introduced the quantum concentration

$$n_Q = \left( \frac{m k_B T}{2\pi \hbar^2} \right)^{\frac{3}{2}}. \tag{4.70}$$

We can relate the quantum concentration to the particle density corresponding to having a single particle in a cube of side length equal to the thermal de Broglie wavelength

$$\lambda_T \sim \frac{\hbar}{\sqrt{2 m k_B T}} \sim \frac{h}{m \tilde{v}}, \tag{4.71}$$

where

$$\tilde{v} = \sqrt{\frac{2\pi k_B T}{m}} \tag{4.72}$$

is a characteristic thermal velocity scale.[3] We will see that the quantity $n_Q$ arises often when discussing quantum statistical mechanics and provides a convenient way to define a demarcation between quantum and classical regimes.

## 4.5 Ideal Gas

In the examples in Section 4.4, we considered single-particle partition functions, but in a gas we need to take into account that there are many particles. Before we can give a statistical mechanical treatment of ideal gases, we need to understand what happens when we have multiple subsystems coupled to a bath.

### 4.5.1 Uncoupled Subsystems

Consider the situation in which a system is composed of two non-interacting (or more realistically, very weakly interacting) subsystems, $A$ and $B$, both of which are connected to a heat bath which has temperature $T$, as illustrated in Fig. 4.5.

---

[3] Another way to see how the thermal de Broglie wavelength arises is to recall that $E = \frac{p^2}{2m}$ and then the corresponding de Broglie wavelength is $\lambda = \frac{h}{p} = \frac{h}{\sqrt{2mE}}$. If we choose $E \sim k_B T$ then up to factors of $2\pi$, which give a particularly simple form for the partition function, the de Broglie wavelength is $\lambda_T$.

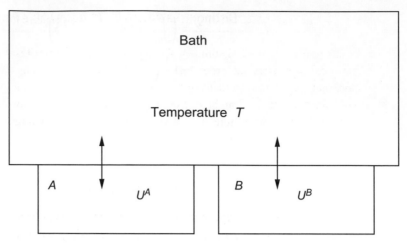

Fig. 4.5  Two non-interacting independent subsystems $A$ and $B$ connected to a bath.

We define the states of subsystem $A$ as $s_i^A$ with energies $\epsilon_i^A$ and the states of subsystem $B$ as $s_j^B$ with energies $\epsilon_j^B$, where $i$ and $j$ label the states of $A$ and $B$, respectively. The states of the system will be pairs $(s_i^A, s_j^B)$, because the lack of interactions means that the state of $A$ does not affect the state of $B$ and vice versa, and the energy of the state will be $\epsilon_i^A + \epsilon_j^B$,[4] leading to a partition function for the whole system of

$$Z = \sum_{ij} e^{-\beta\left(\epsilon_i^A + \epsilon_j^B\right)}$$

$$= \left(\sum_i e^{-\beta\epsilon_i^A}\right)\left(\sum_j e^{-\beta\epsilon_j^B}\right)$$

$$= Z_A Z_B, \tag{4.73}$$

so we see that the partition function factorizes when a system is composed of independent subsystems.

The Helmholtz free energy then takes the form

$$F = -k_B T \ln Z$$

$$= -k_B T \ln(Z_A Z_B)$$

$$= -k_B T \ln(Z_A) - k_B T \ln(Z_B)$$

$$= F_A + F_B, \tag{4.74}$$

which is just the sum of the free energies of the independent subsystems. The same is true for other extensive quantities, such as the entropy and average energy.

---

[4] In general there may be interactions between two subsystems $A$ and $B$, in which case the energy of the state would have the form $\epsilon_i^A + \epsilon_j^B + \epsilon_{int;ij}^{AB}$, where $\epsilon_{int;ij}^{AB}$ is an interaction energy.

## 4.5.2  Distinguishable and Indistinguishable Particles

Whether particles are distinguishable or indistinguishable has consequences for their statistical properties. In order that particles be distinguishable, there needs to be some label that can be used to identify each particle uniquely – for example, particles that sit on fixed sites on a lattice can be distinguished by their position. However, quantum particles such as electrons or protons have no such label and are indistinguishable. We explore the consequences of these differences below.

### Distinguishable Particles

From the result for independent subsystems that we established in Eq. (4.73), we can see that if we have $N$ particles that are non-interacting, we can treat each particle as an independent subsystem and the total partition function is a product of the partition functions for the independent particles. Hence if we know the partition function for a single particle $Z_1$, then the total partition function is

$$Z = (Z_1)^N,  \tag{4.75}$$

if the particles are distinguishable.

We can approach this more formally as follows: if there are $n_s$ particles occupying the energy level $\epsilon_s$, then the total energy for a given set of occupation numbers $\{n_s\}$ is

$$E_{\{n_s\}} = \sum_s n_s \epsilon_s,  \tag{4.76}$$

where the $n_s$ have to satisfy the constraint that

$$N = \sum_s n_s.  \tag{4.77}$$

The partition function is

$$Z = \sum_{\{n_s\}} g_{\{n_s\}} e^{-\beta E_{\{n_s\}}},  \tag{4.78}$$

where $g_{\{n_s\}}$ is the degeneracy associated with the set of occupation numbers $\{n_s\}$, i.e. the number of different ways we can arrange the distinguishable particles between the energy levels for a given $\{n_s\}$, which is given by

$$g_{\{n_s\}} = \frac{N!}{n_0!\, n_1!\ldots}.  \tag{4.79}$$

Then the partition function takes the form

$$Z = \sum_{\{n_s\}} \frac{N!}{\prod_s n_s!} e^{-\beta \sum_s n_s \epsilon_s}$$

$$= \sum_{\{n_s\}} \frac{N!}{\prod_s n_s!} \prod_s e^{-\beta n_s \epsilon_s}$$

$$= \left( \sum_s e^{-\beta \epsilon_s} \right)^N$$

$$= (Z_1)^N, \tag{4.80}$$

where we used the multinomial theorem for $k_1 + k_2 + \cdots = N$:

$$(x_1 + x_2 + \cdots)^N = \sum_{k_1, k_2, \ldots} \frac{N!}{k_1! \, k_2! \ldots} x_1^{k_1} x_2^{k_2} \cdots, \tag{4.81}$$

in going from the second to the third line to obtain Eq. (4.80).

## Indistinguishable Particles

If we consider indistinguishable particles, then for a given set of occupation numbers, there is only one way to arrange the particles. As illustrated in Fig. 4.6, for distinguishable particles we can imagine a label on each of the particles, so the two arrangements of

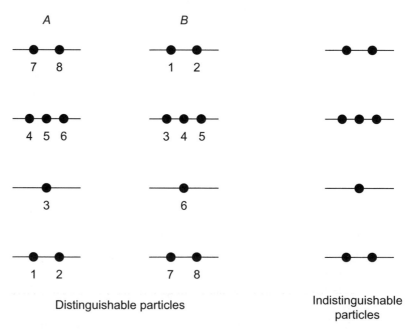

Distinguishable particles                    Indistinguishable particles

**Fig. 4.6**  Two arrangements ($A$ and $B$) of distinguishable particles (labelled 1 to 8) and one arrangement of indistinguishable particles on a set of energy levels. There are clearly multiple distinct ways to arrange the particles when they are distinguishable, but only one way when they are indistinguishable.

the particles, $A$ and $B$, are distinct. If the particles are indistinguishable, then $A$ and $B$ correspond to the same arrangement of particles. When we write down the partition function we get

$$Z = \sum_{\{n_s\}} e^{-\beta E_{\{n_s\}}}, \qquad (4.82)$$

which cannot be factored in the same way as for distinguishable particles.

The difference between distinguishable and indistinguishable particles becomes very important when we deal with quantum particles (see Chapter 7), which in addition to being indistinguishable may have further constraints, depending on whether they are bosonic or fermionic. Fermions have half-integer spin, and common examples include electrons, protons, neutrons, neutrinos and $^3$He, whereas bosons have integer spin and examples include photons, phonons, magnons and $^4$He. An important difference between fermions and bosons is that fermions obey the Pauli exclusion principle, which states that no two fermions can be in the same state.

---

**Example** An example which can help to clarify our discussion is when we have two particles that can occupy three available energy levels (1, 2 and 3). If we assume the particles are distinguishable, there are $9 = 3^2$ microstates for the particles, as written out below:

**Energy level occupied by each particle**

| | | | | | | | | | |
|---|---|---|---|---|---|---|---|---|---|
| **Particle 1**: | 1 | 1 | 1 | 2 | 2 | 2 | 3 | 3 | 3 |
| **Particle 2**: | 1 | 2 | 3 | 1 | 2 | 3 | 1 | 2 | 3 |

For indistinguishable particles, the pairs of occupied levels $\{(1, 2), (2, 1)\}$, $\{(1, 3), (3, 1)\}$ and $\{(2, 3), (3, 2)\}$ are all equivalent, so there are only six possible microstates – this is the number of microstates for indistinguishable bosons. For indistinguishable fermions, the pairs of occupied levels $(1, 1)$, $(2, 2)$ and $(3, 3)$ are forbidden by the Pauli principle, so there are only three possible microstates.

---

We may rewrite the partition function for indistinguishable particles to reflect the differences between fermions and bosons:

$$Z = \sum_{\{n_s\}}^{X} e^{-\beta E_{\{n_s\}}}, \qquad (4.83)$$

where $X$ is the upper limit of allowed $n_s$ values for a level $s$ and takes the value 1 for fermions and $\infty$ for bosons. As before, the constraint $\sum_s n_s = N$ holds for the occupation numbers.

This does not seem to make things any easier. However, there is a limit where it is possible to achieve simpler results. If we look at the limit in which there are many more available energy levels than particles, then the probability of double or higher multiple occupancy of a state is miniscule. In this situation, the distinction between fermions and bosons is essentially irrelevant. This is the classical limit and can be realized when the density $n \ll n_Q$, which corresponds to either the low-density or high-temperature limit in which the interparticle spacing is much larger than the thermal de Broglie wavelength. If there are $N$ particles, then for any microstate, if the particles were distinguishable, there would be $N!$ ways to arrange them, whereas there is only one way to arrange indistinguishable particles, as illustrated in Fig. 4.6. Hence in this limit

$$Z_{\text{indistinguishable}} \simeq \frac{Z_{\text{distinguishable}}}{N!} = \frac{(Z_1)^N}{N!}. \tag{4.84}$$

We will return to examine situations in which the distinctions between fermions and bosons are important later in Chapters 7, 8 and 9.

### 4.5.3 Ideal Gas

#### Monatomic Gas

We can apply our discussion of indistinguishable particles to describe a dilute ideal gas. In an ideal gas, the particles are indistinguishable, so that for $N$ atoms, the partition function is given by Eq. (4.84) and we have already calculated $Z_1$ as the partition function for a particle in a box. If we allow for spin degeneracy (which we didn't in our previous calculation of $Z_1$), then we should multiply $Z_1$ by an extra factor $g_s$ and hence we have

$$Z_1 = g_s n_Q V. \tag{4.85}$$

Using Stirling's formula we can rewrite $N! \simeq \sqrt{2\pi N} N^N e^{-N}$, and hence in the large-$N$ limit when we substitute Eq. (4.85) in Eq. (4.84) we get

$$Z = \frac{1}{\sqrt{2\pi N}} \left( \frac{g_s V e n_Q}{N} \right)^N = \frac{1}{\sqrt{2\pi N}} \left( \frac{g_s e n_Q}{n} \right)^N, \tag{4.86}$$

where $n = N/V$ is the number density of atoms. We can use $Z$ to obtain the Helmholtz free energy as

$$F = -k_B T \ln Z = -N k_B T \ln \left( \frac{g_s n_Q}{n} \right) - N k_B T \ln(e) + \frac{k_B T}{2} \ln(2\pi N), \tag{4.87}$$

and we can drop the last term as it is not extensive, so we have

$$F = -N k_B T \ln \left( \frac{g_s n_Q}{n} \right) - N k_B T. \tag{4.88}$$

From here we can calculate all thermodynamic properties of interest, for instance the pressure

$$P = -\left. \frac{\partial F}{\partial V} \right|_T$$
$$= \frac{N k_B T}{V}, \tag{4.89}$$

so we get the ideal gas law

$$PV = N k_B T, \tag{4.90}$$

which can also be written as

$$PV = nRT. \tag{4.91}$$

The form in Eq. (4.91) follows when we recall that Boltzmann's constant is $k_B = 1.3806 \times 10^{-23}$ J K$^{-1}$ with $N$ the number of molecules and the ideal gas constant is $R = 8.314$ J mol$^{-1}$ K$^{-1}$ with $n$ the number of moles in the gas (1 mole is Avogadro's number $N_A = 6.022 \times 10^{23}$ molecules), so that $Nk_B = nR$. We can also calculate the entropy from

$$S = -\left.\frac{\partial F}{\partial T}\right|_{N,V}$$

$$= Nk_B \ln\left(\frac{g_s n_Q}{n}\right) + Nk_B + \frac{3}{2}Nk_B, \tag{4.92}$$

which gives

$$S = Nk_B \left[\ln\left(\frac{g_s n_Q}{n}\right) + \frac{5}{2}\right], \tag{4.93}$$

which is known as the **Sackur–Tetrode equation** for a monatomic gas, obtained independently by Otto Sackur and Hugo Tetrode in around 1912.

With $F$ and $S$ known, it is straightforward to calculate the energy

$$U = F + TS = \frac{3}{2}Nk_BT, \tag{4.94}$$

and hence the heat capacity at constant volume is

$$C_V = \left.\frac{\partial U}{\partial T}\right|_V = \frac{3}{2}Nk_B. \tag{4.95}$$

We can use Eqs (4.95) and (4.25) to estimate the size of energy fluctuations in an ideal gas and verify our assertion in Section 4.2.3 that in equilibrium, the canonical and microcanonical ensembles are equivalent. From Eqs (4.94) and (4.25) we can estimate the relative size of energy fluctuations:

$$\frac{\Delta U}{U} = \frac{\sqrt{\left(\langle E^2 \rangle - \langle E \rangle^2\right)}}{U}$$

$$= \frac{\sqrt{k_B T^2 C_V}}{U}$$

$$= \frac{\sqrt{k_B T^2 \frac{3}{2} Nk_B}}{\frac{3}{2}Nk_BT}$$

$$= \sqrt{\frac{2}{3N}}. \tag{4.96}$$

For a mole of ideal gas this gives a fractional fluctuation of $\sim 1.05 \times 10^{-12}$, which is too small to be measured.

## Gas of Polyatomic Molecules

For a single polyatomic molecule, there are additional degrees of freedom beyond translations, such as rotations, vibrations and internal electronic transitions. To a good approximation these degrees of freedom can often be treated as independent, so we may write the single-particle partition function as

$$Z_1 = Z_t Z_r Z_v Z_e, \qquad (4.97)$$

where $Z_t$ is the single-particle partition function corresponding to translations that we discussed for the monatomic gas. $Z_r$ is the partition function for rotations: for a linear molecule, such as a diatomic molecule, or a triatomic molecule like $CO_2$, $Z_r$ would be given by the result we found for a linear rotor in Eq. (4.61), $Z_v$ is the vibrational partition function that might be approximated by a simple harmonic oscillator (at low enough temperatures that anharmonicities can be ignored) and $Z_e$ is the partition function for electronic transitions, which might be modelled by a few-level system for low temperatures (relative to the energy spacing of the transitions). An order of magnitude estimate of the characteristic temperatures for these extra degrees of freedom is $\theta_r \sim 10 - 100$ K, $\theta_v \sim 1000$ K and $\theta_e \sim 10000$ K.

Using the expression for $Z_t$ that we obtained in Eq. (4.85), we can follow the steps to get to Eq. (4.86) and see that

$$Z = \frac{1}{\sqrt{2\pi N}} \left( \frac{g_s e n_Q}{n} \right)^N Z_r^N Z_v^N Z_e^N, \qquad (4.98)$$

and the Helmholtz free energy is

$$F = -Nk_B T \ln \left( \frac{g_s n_Q}{n} \right) - Nk_B T - Nk_B T \ln(Z_r) - Nk_B T \ln(Z_v) - Nk_B T \ln(Z_e). \quad (4.99)$$

Now, $Z_r$, $Z_v$ and $Z_e$ are all independent of volume, so we still obtain the ideal gas law

$$PV = Nk_B T \qquad (4.100)$$

when we take a derivative of $F$ with respect to $V$ to calculate the pressure. If we are interested in temperatures much less than $\theta_v$ and $\theta_e$ (which is the case at room temperature, for example) then we can neglect the terms above involving $Z_v$ and $Z_e$ and approximate the entropy as

$$S = Nk_B \left[ \ln \left( \frac{g_s n_Q}{n} \right) + \frac{5}{2} \right] + Nk_B \frac{\partial}{\partial T} [T \ln Z_r].$$

At temperatures $T \gg \theta_r$ we know from Eq. (4.64) that $Z_r \simeq \frac{T}{\theta_r}$, in which case we get (neglecting vibrations and electronic excitations)

$$S = Nk_B \left[ \ln \left( \frac{g_s n_Q}{n} \frac{T}{\theta_r} \right) + \frac{7}{2} \right], \qquad (4.101)$$

and hence the energy is

$$U = \frac{5}{2} Nk_B T, \qquad (4.102)$$

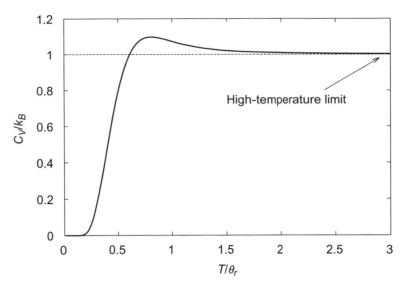

Fig. 4.7 Heat capacity from rotations for a gas of diatomic molecules – note the small peak at a temperature $T \sim 0.8\,\theta_r$.

which implies a heat capacity of

$$C_V = \frac{5}{2}Nk_B. \tag{4.103}$$

At temperatures much below $\theta_r$, we know that $Z_r \simeq 1 + 3\exp\{-2\theta_r/T\}$, which will lead to an exponentially small contribution to the heat capacity so that $C_V = \frac{3}{2}Nk_B$ as for a monatomic gas. The contribution to the heat capacity from rotations over the full temperature range is plotted in Fig. 4.7.

### 4.5.4  Example: Entropy of Mixing

The factor of $N!$ in Eq. (4.84) has important physical consequences, as first recognized by Gibbs. This is most easily seen in the example of the entropy of mixing. Suppose we have a container with a partition that separates two regions, each with volume $V$. Consider two scenarios: in scenario 1, both sides are filled with an ideal gas containing $N$ molecules of type $A$, whereas in scenario 2, one side is filled with an ideal gas containing $N$ molecules of type $A$ and the other contains $N$ molecules of type $B$, as illustrated in Fig. 4.8. The partition is then removed and then we wait for the gas to equilibrate. We can calculate the change in entropy in each scenario, using the Sackur–Tetrode equation, Eq. (4.93).

In scenario 1, the initial entropy is given by twice the entropy for $N$ particles in volume $V$:

$$S_1^{\text{initial}} = 2Nk_B\left[\ln\left(\frac{g_s^A n_Q^A V}{N}\right) + \frac{5}{2}\right], \tag{4.104}$$

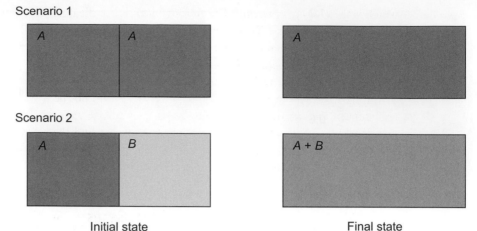

Scenario 1

Scenario 2

Initial state                                        Final state

**Fig. 4.8**    Two scenarios for entropy of mixing. In scenario 1 there are atoms of type $A$ on both sides of a partition before it is removed and in scenario 2 there are atoms of type $A$ on one side of the partition and atoms of type $B$ on the other side of the partition before it is removed.

where $g_s^A$ is the spin degeneracy and $n_Q^A$ is the quantum concentration for molecules of type $A$. The final entropy is that of $2N$ particles in a volume $2V$:

$$S_1^{\text{final}} = 2Nk_B \left[ \ln \left( \frac{g_s^A n_Q^A 2V}{2N} \right) + \frac{5}{2} \right]. \tag{4.105}$$

Hence, the change in entropy is

$$\Delta S_1 = S_1^{\text{final}} - S_1^{\text{initial}} = 0. \tag{4.106}$$

This is in accord with our expectation that the process of removing the partition is reversible. If we reinsert the partition, then we are back to our original situation in which there are $N$ molecules of type $A$ on each side of the partition – they may not be the same molecules as were there originally, but since they are indistinguishable the situation is equivalent to the original configuration.

In scenario 2, the initial entropy is that of $N$ molecules in volume $V$ for both types of molecules:

$$S_2^{\text{initial}} = Nk_B \left[ \ln \left( \frac{g_s^A n_Q^A V}{N} \right) + \frac{5}{2} \right] + Nk_B \left[ \ln \left( \frac{g_s^B n_Q^B V}{N} \right) + \frac{5}{2} \right], \tag{4.107}$$

while the final entropy is that of $N$ molecules in volume $2V$ for both types of molecules:

$$S_2^{\text{final}} = Nk_B \left[ \ln \left( \frac{g_s^A n_Q^A 2V}{N} \right) + \frac{5}{2} \right] + Nk_B \left[ \ln \left( \frac{g_s^B n_Q^B 2V}{N} \right) + \frac{5}{2} \right]. \tag{4.108}$$

Hence the change in entropy is

$$\Delta S_2 = S_2^{\text{final}} - S_2^{\text{initial}} = 2Nk_B \ln 2, \tag{4.109}$$

which is the *entropy of mixing*. The entropy increases, indicating that scenario 2 describes an irreversible process. This is in accord with the observation that if we replace the partition, then the situation is not the same as it was initially – there will be molecules of both type $A$ and type $B$ on both sides of the partition.

If we review the derivation of the Sackur–Tetrode equation then we see that if the factor of $N!$ had not been included in Eq. (4.84), there would not be a factor of $N$ inside the argument of the logarithm. This would imply that $\Delta S_1 = \Delta S_2$, even though, as we have argued above, one corresponds to a reversible process and the other corresponds to an irreversible process. We will see additional arguments for the presence of $N!$ in the partition function for dilute indistinguishable particles in Chapter 7.

## 4.6 Non-ideal Gases

We have explored the properties of ideal gases in considerable detail, but ignoring interactions between molecules is an oversimplification. For a more quantitative description of real gases it is necessary to include interactions. We discuss these interactions and show that a simple treatment leads to the van der Waals equation of state.

We consider $N$ classical particles, with the momentum and position of the $i^{\text{th}}$ particle labelled $\mathbf{p}_i$ and $\mathbf{r}_i$, respectively. We assume that the particles located at $\mathbf{r}_i$ and $\mathbf{r}_j$ have pairwise interactions via a potential $\phi(|\mathbf{r}_i - \mathbf{r}_j|)$ that depends only on the distance between the particles, $r_{ij} = |\mathbf{r}_i - \mathbf{r}_j|$. The energy of the system is thus given by

$$E = \sum_i \frac{p_i^2}{2m} + \sum_{j>i} \phi(r_{ij}), \tag{4.110}$$

where the sum is for $j > i$ in the interaction term so that we do not double count any interactions.

A typical interaction potential has the characteristics embodied by the Lennard–Jones potential

$$\phi_{\text{LJ}}(r) = 4\epsilon \left\{ \left(\frac{r_0}{r}\right)^{12} - \left(\frac{r_0}{r}\right)^6 \right\}, \tag{4.111}$$

which is strongly repulsive at short distances ($r < r_0$) and weakly attractive at long distances, as illustrated in Fig. 4.9.[5]

We can write down the $N$-particle classical partition function

$$Z_N = \frac{g_s^N}{N!} \int \frac{d^3\mathbf{p}_1 \dots d^3\mathbf{p}_N d^3\mathbf{r}_1 \dots d^3\mathbf{r}_N}{(2\pi\hbar)^{3N}} e^{-\beta E}, \tag{4.112}$$

and substituting Eq. (4.110) for the energy $E$ we get

$$Z_N = \frac{g_s^N}{N!} \int \frac{d^3\mathbf{p}_1 \dots d^3\mathbf{p}_N}{(2\pi\hbar)^{3N}} e^{-\beta \sum_i \frac{p_i^2}{2m}} \int d^3\mathbf{r}_1 \dots d^3\mathbf{r}_N e^{-\beta \sum_{j>i} \phi(r_{ij})}. \tag{4.113}$$

---

[5] The Lennard–Jones potential is widely used in simulations of fluids as it captures qualitatively important aspects of more realistic interaction potentials.

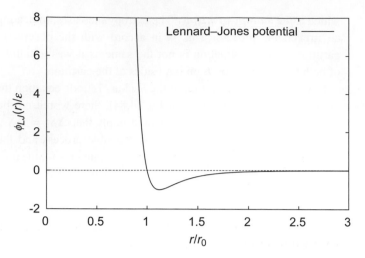

**Fig. 4.9** Lennard–Jones potential: note the strongly repulsive short-range piece for $r < r_0$ and the weakly attractive tail for $r > r_0$.

We can perform the momentum integrals to get

$$\left[ \int \frac{d^3\mathbf{p}}{(2\pi\hbar)^3} e^{-\frac{\beta p^2}{2m}} \right]^N = \left( \frac{mk_BT}{2\pi\hbar^2} \right)^{\frac{3N}{2}} = n_Q^N, \tag{4.114}$$

and hence

$$Z_N = \frac{g_s^N n_Q^N}{N!} Z_\phi, \tag{4.115}$$

where

$$Z_\phi = \int d^3\mathbf{r}_1 \dots d^3\mathbf{r}_N e^{-\beta \sum_{j>i} \phi(r_{ij})}. \tag{4.116}$$

Now, if $\phi = 0$, we see that $Z_\phi = V^N$ and hence

$$Z_N = \frac{(Z_1)^N}{N!}, \qquad \text{with} \quad Z_1 = g_s n_Q V, \tag{4.117}$$

just as we already found for an ideal gas.

In general, $Z_\phi$ is non-trivial to evaluate. To obtain an approximate form, assume that we can write $\phi(r)$ as the sum of a short-range repulsive piece $\phi_{sr}(r)$ and a long-range attractive piece $\phi_{lr}(r)$, similar to the Lennard–Jones potential introduced in Eq. (4.111):

$$\phi(r) = \phi_{sr}(r) + \phi_{lr}(r).$$

In this case

$$Z_\phi = \int d^3\mathbf{r}_1 \dots d^3\mathbf{r}_N e^{-\beta \sum_{j>i} [\phi_{sr}(r_{ij})+\phi_{lr}(r_{ij})]}, \tag{4.118}$$

and to evaluate this we make the additional approximation for the long-range part of the potential

$$\sum_{j>i} \phi_{\text{lr}}(r_{ij}) = \frac{1}{2} \sum_i \sum_{j\neq i} \phi_{\text{lr}}(r_{ij})$$

$$\simeq \frac{1}{2} \sum_i \frac{(N-1)}{V} \int d^3\mathbf{r}\, \phi_{\text{lr}}(r), \qquad (4.119)$$

where we replaced the sum over the potentials of the $N-1$ particles $j \neq i$ by $(N-1)$ times the average value of the potential, which is[6]

$$\frac{1}{V} \int d^3\mathbf{r}\, \phi_{\text{lr}}(r).$$

All $N$ contributions to the sum over $i$ will be identical in Eq. (4.119), so in the limit of large $N$ we approximate

$$\sum_{j>i} \phi_{\text{lr}}(r_{ij}) \simeq -\frac{aN^2}{V}, \qquad (4.120)$$

where

$$a = -\frac{1}{2} \int d^3\mathbf{r}\, \phi_{\text{lr}}(r). \qquad (4.121)$$

Thus we can simplify $Z_\phi$ to

$$Z_\phi \simeq e^{\frac{\beta a N^2}{V}} \int d^3\mathbf{r}_1 \ldots d^3\mathbf{r}_N\, e^{-\beta \sum_{j>i} \phi_{\text{sr}}(r_{ij})}. \qquad (4.122)$$

Now, for the short-range potential, $\phi_{\text{sr}} \simeq 0$ almost everywhere except when two particles are very close. The repulsive aspect of the short-range potential will have the effect of excluding all of the other $N-1$ particles in some volume around the $i^{\text{th}}$ particle. Hence, when $N \gg 1$ we can approximate

$$\int d^3\mathbf{r}_1 \ldots d^3\mathbf{r}_N\, e^{-\beta \sum_{j>i} \phi_{\text{sr}}(r_{ij})} \simeq (V - Nb)^N, \qquad (4.123)$$

where $b$ is an average effective volume that is excluded by each molecule, leaving the volume that a molecule can actually move around in as effectively $V - Nb$. With these approximations we obtain the partition function

$$Z \simeq \frac{g_s^N}{N!} \left( \frac{mk_BT}{2\pi\hbar^2} \right)^{\frac{3N}{2}} (V - Nb)^N\, e^{\frac{\beta a N^2}{V}}, \qquad (4.124)$$

where the gas is characterized by the parameters $a$ and $b$. From this approximate partition function it is straightforward (see Problem 4.10) to derive the van der Waals equation of state for a non-ideal gas:

---

[6] We saw in Chapter 1 that this approach is valid for random variables due to the central limit theorem, however, here the long-range potential will induce correlations between the particles so that their positions are not truly random variables, and so this is only an approximation.

$$P = \frac{Nk_BT}{V - Nb} - \frac{aN^2}{V^2}. \tag{4.125}$$

The van der Waals equation of state highlights physics left out in the derivation of the ideal gas law – the finite size of molecules and their weak long-range attractions – and while more accurate than the ideal gas law is not generally adequate to give a quantitative description of real gases.

## 4.7 The Equipartition Theorem

A result that emerges from studying the ideal gas is that the energy is very simple in the classical limit – for the monatomic ideal gas the energy is $\frac{3}{2}Nk_BT$, and in the diatomic gas at temperatures high enough that all rotations are accessible the energy is $\frac{5}{2}Nk_BT$. We notice that in the monatomic gas, the energy is purely kinetic, and there are contributions from motion in the $x$, $y$ and $z$ directions, which average out to $\frac{1}{2}k_BT$ per particle per degree of freedom. When we allow rotations for a diatomic molecule, we gain two more factors of $\frac{1}{2}k_BT$ per particle in the classical limit. We note that there are only two non-zero moments of inertia for a diatomic molecule, hence there are two degrees of freedom, as we might have guessed from counting multiples of $\frac{1}{2}k_BT$.

The results noted above seem to indicate a pattern, and indeed they can be stated more formally in the **equipartition theorem**:

> *For a classical system in equilibrium, the average energy of each degree of freedom that the energy is quadratic in is $\frac{1}{2}k_BT$.*

To prove this result, suppose that the energy of a system with $N$ quadratic degrees of freedom can be written as

$$E(\{x_i\}) = \sum_{i=1}^{N} \epsilon_i = \sum_{i=1}^{N} c_i x_i^2, \tag{4.126}$$

where $x_i$ refers to any co-ordinate of the system (which could be position, momentum, angular momentum, etc.) and the $\{c_i\}$ are a set of constants. Now, the mean energy

$$U = \langle E \rangle = \left\langle \sum_{i=1}^{N} \epsilon_i \right\rangle = \sum_{i=1}^{N} \langle \epsilon_i \rangle \tag{4.127}$$

is a sum over the mean energies of each of the co-ordinates, and

$$\langle \epsilon_j \rangle = \int d\Gamma \, \epsilon_j \, P(\epsilon_j), \tag{4.128}$$

where $d\Gamma = \prod_{i=1}^{N} dx_i$. In the canonical ensemble we get

$$\langle \epsilon_j \rangle = \frac{\int d\Gamma \, \epsilon_j \, e^{-\beta E(\{x_i\})}}{\int d\Gamma e^{-\beta E(\{x_i\})}}$$

$$
\begin{aligned}
&= \frac{\int d\Gamma\, \epsilon_j\, e^{-\beta \sum_l \epsilon_i}}{\int d\Gamma\, e^{-\beta \sum_i \epsilon_i}} \\
&= \frac{\int dx_j\, \epsilon_j e^{-\beta \epsilon_j} \int d\Gamma'\, e^{-\beta \sum_{i \neq j} \epsilon_i}}{\int dx_j\, e^{-\beta \epsilon_j} \int d\Gamma'\, e^{-\beta \sum_{i \neq j} \epsilon_i}},
\end{aligned} \tag{4.129}
$$

where $d\Gamma' = \prod_{i \neq j}^{N} dx_i$. Hence, noting that the integrals over $d\Gamma'$ cancel, we find

$$
\begin{aligned}
\langle \epsilon_j \rangle &= \frac{\int dx_j\, \epsilon_j e^{-\beta \epsilon_j}}{\int dx_j\, e^{-\beta \epsilon_j}} \\
&= \frac{c_j \int_{-\infty}^{\infty} dx_j\, x_j^2 e^{-\beta c_j x_j^2}}{\int_{-\infty}^{\infty} dx_j\, e^{-\beta c_j x_j^2}} \\
&= \frac{c_j \dfrac{\sqrt{\pi}}{2(\beta c_j)^{\frac{3}{2}}}}{\sqrt{\dfrac{\pi}{\beta c_j}}} \\
&= \frac{1}{2\beta} \\
&= \frac{1}{2} k_B T,
\end{aligned} \tag{4.130}
$$

where we used previously established results for Gaussian integrals. The significance of this result is that if we have a system where the energy is quadratic in some variables, then we know their contribution to the energy at high temperatures without having to do any calculation.

### 4.7.1 Example: The Ideal Gas

For an ideal gas, the energy of a single molecule is

$$
E = \frac{p_x^2}{2m} + \frac{p_y^2}{2m} + \frac{p_z^2}{2m}. \tag{4.131}
$$

The energy is quadratic in three variables, $p_x$, $p_y$ and $p_z$, so the equipartition theorem tells us that the energy per particle is

$$
\frac{E}{N} = \frac{1}{2} k_B T + \frac{1}{2} k_B T + \frac{1}{2} k_B T = \frac{3}{2} k_B T, \tag{4.132}
$$

which we have already obtained from a longer calculation.

### 4.7.2 Dulong and Petit Law

In addition to the ideal gas case considered above, another simple application of the equipartition theorem is to simple harmonic oscillators, where the energy is quadratic in position and momentum. We found that for a single harmonic oscillator

$$U = \frac{\hbar\omega}{2} \coth\left(\frac{\hbar\omega}{2k_B T}\right),$$ (4.133)

which in the high-temperature limit, $k_B T \gg \hbar\omega$, becomes

$$U \to \frac{\hbar\omega}{2}\frac{2k_B T}{\hbar\omega} = k_B T,$$ (4.134)

as we would expect from the equipartition theorem for two degrees of freedom.

The Einstein model of a crystal treats the vibrations of a solid (i.e. phonons) as arising from a collection of $3N$ independent harmonic oscillators with the same frequency $\omega$, with the energy given by

$$E = \sum_{i=1}^{3N} \frac{p_i^2}{2m} + \sum_{i=1}^{3N} \frac{1}{2}m\omega^2 x_i^2.$$ (4.135)

Hence at high temperatures (room temperature is usually sufficient) the equipartition theorem gives

$$U = \langle E \rangle = 3N\frac{1}{2}k_B T + 3N\frac{1}{2}k_B T = 3Nk_B T,$$ (4.136)

and hence

$$C_V = \left.\frac{\partial U}{\partial T}\right|_V = 3Nk_B.$$ (4.137)

For a mole of a material, i.e. $N = N_{\text{Avogadro}} = 6.022 \times 10^{23}$, $C_V = 3R$, where $R = 8.314 \text{ J K}^{-1} \text{mol}^{-1}$ is the universal gas constant.

This provides a microscopic explanation for the Dulong and Petit law that the high-temperature limit of the molar heat capacity of many different crystals is $3R$. The law (not stated in the modern form) was proposed by Pierre Louis Dulong and Alex Thérèse Petit in 1819 based on empirical observations. We will obtain a more complete understanding of the heat capacity of solids when we discuss the Debye model in Chapter 9.

## 4.8 Summary

In the canonical ensemble, the temperature $T$, particle number $N$ and volume $V$ are fixed. The equation

$$F = -k_B T \ln Z$$ (4.138)

is the bridge equation which relates a thermodynamic concept, the Helmholtz free energy $F = U - TS$, to a statistical mechanical concept, the partition function, which can be written as a sum over states $s$:

$$Z = \sum_s e^{-\beta\epsilon_s},$$ (4.139)

or a sum over different energy levels $\epsilon_j$, weighted by their degeneracy $g_j$:

$$Z = \sum_j g_j e^{-\beta \epsilon_j}. \tag{4.140}$$

For classical systems the partition function is expressed as an integral over phase space. For multi-particle systems the form of the partition function depends on whether particles are distinguishable or indistinguishable. For distinguishable particles the $N$-particle partition function is

$$Z_N = (Z_1)^N, \tag{4.141}$$

where $Z_1$ is the single-particle partition function and for indistinguishable particles the $N$-particle partition function is only simple in the dilute limit, in which case it takes the form

$$Z_N \simeq \frac{(Z_1)^N}{N!}. \tag{4.142}$$

Treating an ideal gas as the dilute limit of a gas of indistinguishable particles in a box, for which $Z_1 = g_s n_Q V$, with

$$n_Q = \left(\frac{mk_BT}{2\pi\hbar^2}\right)^{\frac{3}{2}}, \tag{4.143}$$

the quantum concentration, one can derive the ideal gas law

$$PV = Nk_BT, \tag{4.144}$$

and that the energy of an ideal gas is

$$U = \frac{3}{2}Nk_BT. \tag{4.145}$$

The result for the energy can also be obtained from the equipartition theorem, which states that for a classical system in equilibrium, each quadratic degree of freedom contributes $\frac{1}{2}k_BT$ to the energy.

# Problems

**4.1** We can derive the probability that a system is in microstate $j$ in the canonical ensemble by considering the system + reservoir as an isolated system with total energy $U$, where the reservoir energy is much larger than that of the system. Suppose the system is in a definite microstate $j$ which has energy $\epsilon_j$.

(a) Justify

$$p_j = c\,\Omega_R(U - \epsilon_j),$$

where $c$ is a constant and $\Omega_R$ is the multiplicity function for the reservoir.

(b) Expand $\ln \Omega_R(U - \epsilon_j)$ to show that

$$p_j = \frac{e^{-\beta\epsilon_j}}{Z},$$

where $\beta = 1/k_B T$ and give an expression for $Z$ in terms of $c$ and $\Omega_R$.

**4.2** We have seen how to calculate $C_V$ for an ideal gas. Calculate $C_P$ for an ideal gas in three dimensions, recalling that it can be written as a derivative of the enthalpy $H$ via

$$C_P = \left.\frac{\partial H}{\partial T}\right|_P = \left.T\frac{\partial S}{\partial T}\right|_P.$$

Relate $C_P$ and $C_V$ for both a monatomic and a diatomic ideal gas.

**4.3** Estimate $\theta_r$ for $O_2$ and $N_2$. You may find it helpful to note that the bond length for oxygen is 1.208 Å and the bond length for nitrogen is 1.0975 Å, and the atomic masses of oxygen and nitrogen are 15.999 amu and 14.0067 amu, respectively.

**4.4** We saw in Eq. (4.74) that for a system composed of two independent subsystems $A$ and $B$, the Helmholtz free energy takes the form

$$F = F_A + F_B,$$

which is just the sum of the free energies of the independent subsystems. Show that the same is true for other extensive quantities such as the entropy and average energy.

**4.5** In Section 4.2.1 we found an expression for the relative probabilities of states with different energies. We can find the same result in a different way:

(a) The ratio of the probabilities of the system having energies $\epsilon_1$ and $\epsilon_2$ can be related to the number of states that the reservoir has with energies $U - \epsilon_1$ and $U - \epsilon_2$. Use an argument of this form to show that the ratio of probabilities is

$$\frac{P(\epsilon_1)}{P(\epsilon_2)} = e^{\frac{1}{k_B}[S(U-\epsilon_1)-S(U-\epsilon_2)]}.$$

(b) Taylor expand the right-hand side of the expression in part (a) to show that

$$\frac{p(\epsilon_1)}{p(\epsilon_2)} = \exp\left\{-\frac{(\epsilon_1 - \epsilon_2)}{k_B T}\right\}.$$

(c) We showed that the relative fluctuations in the energy vanish in the limit $N \to 0$. Show that the probability that the system has an energy $E = \langle E \rangle + \delta E$ is a Gaussian distribution for small $\delta E$.

**4.6** Consider $N$ ionized atoms of a monatomic gas with charge $q$ and mass $m$ placed in a cube of side length $L$, held at temperature $T$ and subject to a uniform electric field $\mathcal{E}$ in the $x$ direction. Ignore interactions between the ions.

(a) Treat the ions as a classical gas with energies

$$E = \frac{p^2}{2m} - q\mathcal{E}x,$$

and let $x \in [-L/2, L/2]$. Calculate the one-particle partition function $Z_1$ and hence show that the free energy for such a gas takes the form

$$F = -Nk_BT \ln\left[\frac{n_Q}{n}\frac{L_\phi}{L} \sinh\left(\frac{L}{L_\phi}\right)\right] - Nk_BT,$$

where $n$ is the concentration and

$$n_Q = \left(\frac{mk_BT}{2\pi\hbar^2}\right)^{\frac{3}{2}}$$

is the quantum concentration. Specify $L_\phi$ in terms of $\beta$, $q$ and $\mathcal{E}$.

(b) Using your answer to part (a) or otherwise, show that the average energy $U$ of the gas is

$$U = Nk_BT\left[\frac{5}{2} - \frac{L}{L_\phi}\coth\left(\frac{L}{L_\phi}\right)\right].$$

(c) Find the limiting forms of the expression for $U$ in part (b) in the high-temperature $(k_BT \gg q\mathcal{E}L)$ and low-temperature $(k_BT \ll q\mathcal{E}L)$ limits and comment briefly on your results.

**4.7** Use the equipartition theorem to estimate the typical angular velocity of a nitrogen molecule at room temperature. (The N–N bond length is 1.0975 Å and the mass of an N atom is 14.0067 amu.)

**4.8** Consider ultra-cold dilute neutrons, at temperature $T = 1$ mK, which are initially placed at the bottom of a box of height $H$. How high does the box need to be to ensure that a neutron with average energy remains in the box?

**4.9** Consider a diatomic ideal gas with $N$ particles in three dimensions at a temperature $T$, where vibrations and internal electronic transitions are unimportant. Calculate the heat capacity at constant volume, $C_V$, by first finding the partition function for the gas. You may assume that the single-particle partition function for rotations is

$$Z_{\text{rotations}} \simeq \frac{T}{\theta_r},$$

where $\theta_r$ is a characteristic temperature for rotations. How would your result change if you considered a monatomic ideal gas?

**4.10** In an ideal gas the particles are assumed to be non-interacting. For some real gases, a reasonable approximation to the canonical partition function for a gas with $N$ particles at temperature $T$ in volume $V$ is

$$Z = c\left(\frac{V - Nb}{N}\right)^N \left(\frac{2\pi mk_BT}{\hbar^2}\right)^{\frac{3N}{2}} e^{\frac{N^2 a}{Vk_BT}},$$

where $a$, $b$ and $c$ are positive constants. Show that this partition function implies the van der Waals equation of state

$$(P + n^2 a)(1 - nb) = \frac{Nk_BT}{V},$$

where $n = N/V$.

**4.11** Suppose we have a lattice of $N$ indistinguishable atoms ($N \gg 1$) in one of three states with energies $0$, $\epsilon$ and $4\epsilon$.

(a) If the mean energy per atom is $\epsilon$, what is the temperature?
(b) Find the entropy as a function of temperature at this energy.

**4.12** Consider dilute He gas in a three-dimensional box with the initial condition that all atoms have a speed of $1000 \text{ m s}^{-1}$ but are moving in random directions. After some time the gas will equilibrate. What will be the temperature of the gas at equilibrium (neglect the exchange of heat between the gas and the environment)?

**4.13** Calculate the entropy $S$ and energy $U$ for a quantum harmonic oscillator and hence determine the heat capacity at constant volume, $C_V$.

**4.14** Calculate the entropy $S$ and energy $U$ for a classical harmonic oscillator and hence determine the heat capacity at constant volume, $C_V$. Comment on whether your answer agrees with what you expect from the equipartition theorem.

**4.15** Suppose that a system of $N$ atoms of type $A$ is placed in diffusive contact with a system of $N$ atoms of type $B$ at the same temperature and volume. Show that after diffusive equilibrium is reached, the total entropy has increased by $2N \ln 2$ compared to the entropy when the systems were not in contact. This entropy increase is known as the entropy of mixing. If the atoms are identical ($A = B$), show that there is no increase in entropy when diffusive contact is established.

**4.16** Find the partition function for a symmetric rotor in the classical limit. Hence compare the heat capacity of a gas of molecules that can be modelled as symmetric rotors with a gas of molecules that can be modelled as linear rotors. Discuss your result with reference to the equipartition theorem.

**4.17** Consider non-interacting spins as we did in Section 2.4. The energy for each spin $s$ is given by

$$\epsilon = -s\mu B,$$

where the spin $s$ can take values $\pm 1$. For a system of $N$ spins in a solid, find the Helmholtz free energy.

(a) Using the result that the magnetization is

$$M = -\frac{1}{\mu}\frac{\partial F}{\partial B},$$

find the magnetization per spin, $m$, and show that it agrees with the result we found in Eq. (2.36).
(b) Find the heat capacity at constant magnetic field and show that it displays a Schottky anomaly as a function of temperature.

**4.18** Deuterium hydride (HD) has a boiling point of 22.1 K and characteristic temperatures $\theta_r = 64$ K and $\theta_v = 5518$ K for rotations and vibrations, respectively. At atmospheric pressure and a temperature of 3000 K the gas starts to dissociate.

Treating HD molecules as a linear rotor, make a plot of the heat capacity over the temperature range 22.1 K to 3000 K.

**4.19** Hydrogen gas ($H_2$) can exist in two slightly different forms, ortho and para hydrogen, depending on the state of the spins of the hydrogen nuclei. Treating hydrogen as a linear rotor, the rotational energy is

$$\epsilon_J = \frac{\hbar^2 J(J+1)}{2I}.$$

For ortho hydrogen the spins are in a triplet state, for which the degeneracy is 3 and $J$ can only take even integer values $0, 2, 4, \ldots$, while for para hydrogen the spins are in a singlet state, for which the degeneracy is 1 and $J$ can only take odd integer values $1, 3, 5, \ldots$. Noting that the proton–proton bond distance in hydrogen is 0.747 Å, find the ratio of ortho to para hydrogen at (i) a temperature of 25 K and (ii) high temperatures.

You may ignore non-rotational degrees of freedom since these are the same for both species.

**4.20** Solve Problem 1.9 using the canonical ensemble by recalling that the energy of the polymer chain is given by

$$E = \sum_i \frac{p_i^2}{2m} - \mathbf{X} \cdot \mathbf{R},$$

where $\mathbf{X}$ is the tension applied to the chain and $\mathbf{R} = \sum_i \mathbf{r}_i$ is the extension of the chain. Parametrize the orientation of the $i^{\text{th}}$ monomer by a vector written in spherical polar co-ordinates:

$$\mathbf{r}_i = a\left(\cos\phi_i \sin\theta_i, \sin\phi_i \sin\theta_i, \cos\theta_i\right).$$

Assuming that $\mathbf{X}$ is oriented along the $z$ direction, show that the $z$ component of the average displacement is

$$\langle R_z \rangle = \frac{Na}{\sinh(\beta X a)}\left[\cosh(\beta X a) - \frac{\sinh(\beta X a)}{\beta X a}\right].$$

Specify under what circumstances we may consider the tension as small, and show that in this limit the polymer may be regarded as a spring with spring constant

$$k = \frac{3k_B T}{Na^2}.$$

**4.21** Consider an ideal gas confined in a potential

$$V(x, y, z) = \alpha(x^n + y^n + z^n),$$

where $n$ is a positive even integer. Calculate the classical partition function and hence show that the heat capacity is

$$C_V = \left(\frac{3}{2} + \frac{3}{n}\right)Nk_B.$$

# Kinetic Theory

Particles suspended in a fluid move in an apparently random fashion in what is known as Brownian motion (named after the botanist Robert Brown, who observed the motion of pollen grains in water in 1827). This can be understood by treating the fluid as composed of many molecules and then treating the motion of these molecules in the fluid statistically. In most of our discussions we have related statistical mechanics to thermodynamics, by using microscopic degrees of freedom to calculate some statistical quantity, e.g. $\Omega$ in the microcanonical ensemble, or $Z$ in the canonical ensemble, and then connected these quantities to thermodynamic variables, such as entropy or the Helmholtz free energy. Once we have made a connection via a bridge equation, the nature of the microscopic degrees of freedom is irrelevant for thermodynamic behaviour. In kinetic theory, we instead work with the microscopic degrees of freedom directly to calculate macroscopic quantities. Kinetic theory also gives a means to calculate transport properties, such as conductivities or diffusion coefficients. We first apply kinetic theory to the familiar system of a classical ideal gas, which allows us to view it from a different point of view to the one we have taken thus far, that it is the classical limit of $N$ non-interacting particles confined in a box. To do this we will derive the distribution of particle velocities in an ideal gas, the Maxwell–Boltzmann distribution. We then use a simple application of kinetic theory to obtain the average pressure and the effusion rate for an ideal gas. We finish with a discussion of Brownian motion and diffusion.

## 5.1 Maxwell–Boltzmann Velocity Distribution

We know from the canonical ensemble that for an ideal gas the probability that the system is in any particular microstate with energy between $E$ and $E + dE$ and partition function $Z$ is given by

$$\frac{e^{-\beta E}}{Z}.$$

As we discussed in Section 4.5.3, we can understand the properties of an ideal gas by considering a single particle and then obtain the partition function of a many-particle system from the single-particle partition function Eq. (4.85):

$$Z_1 = g_s n_Q V = g_s V \left( \frac{m k_B T}{2\pi\hbar^2} \right)^{\frac{3}{2}}. \tag{5.1}$$

In order to find the probability that the system has energy $E$, we need to know how many microstates there are that have energy within the energy range $[E, E + dE]$, which we calculate below.

## 5.1.1  Density of States

To obtain the density of states, we first pick an energy $E^*$ and then calculate the number of states with energy $E < E^*$. For particle in a box states in three dimensions, for a cubic box with side length $L$ and volume $V = L^3$, we can write the energy eigenvalues as

$$E = \frac{\hbar^2}{2m} \frac{\pi^2}{V^{\frac{2}{3}}} \left( n_x^2 + n_y^2 + n_z^2 \right), \tag{5.2}$$

where $n_x$, $n_y$ and $n_z$ are positive integers, so we require

$$n_x^2 + n_y^2 + n_z^2 < \left( \frac{2mE^*}{\hbar^2} \right) \frac{V^{\frac{2}{3}}}{\pi^2}, \tag{5.3}$$

which represents 1/8 of a sphere (in energy space) with radius $r = \sqrt{\frac{2mE^*}{\hbar^2}} \frac{V^{\frac{1}{3}}}{\pi}$. Now, for a sphere of radius $r$, the number of points $(n_x, n_y, n_z)$ in the octant of the sphere, where $n_x, n_y, n_z > 0$, is $\frac{1}{8} \times \frac{4}{3}\pi r^3 = \frac{1}{6}\pi r^3$, so the number of states in the relevant portion of the sphere with energy $\leq E$ is (including a factor $g_s$ to take into account spin degeneracy) is

$$N(E) = \frac{g_s \pi}{6} \left( \frac{2mE}{\hbar^2} \right)^{\frac{3}{2}} \frac{V}{\pi^3} = \frac{g_s}{6\pi^2} \left( \frac{2m}{\hbar^2} \right)^{\frac{3}{2}} V E^{\frac{3}{2}}. \tag{5.4}$$

If we look in the energy range $[E, E + dE]$, then the number of states is

$$N(E + dE) - N(E) \simeq \frac{dN}{dE} dE$$

$$= \frac{g_s}{4\pi^2} \left( \frac{2m}{\hbar^2} \right)^{\frac{3}{2}} V E^{\frac{1}{2}} dE. \tag{5.5}$$

Thus the density of states, i.e. the number of states per unit energy (which we can also view as the degeneracy of states at energy $E$) is

$$g(E) = \frac{dN}{dE} = \frac{g_s}{4\pi^2} \left( \frac{2m}{\hbar^2} \right)^{\frac{3}{2}} V E^{\frac{1}{2}}. \tag{5.6}$$

We discuss an alternative method to calculate the density of states in Chapter 7.

## 5.1.2  Maxwell–Boltzmann Velocity Distribution

Armed with the density of states, we can calculate the probability density that the system has energy $E$ by noting that the probability that the energy lies between $E$ and $E + dE$ is

$$p(E)dE = g(E)\frac{e^{-\beta E}}{Z}dE$$

$$= \frac{g_s}{4\pi^2}\left(\frac{2m}{\hbar^2}\right)^{\frac{3}{2}}VE^{\frac{1}{2}}\frac{1}{g_sV\left(\frac{mk_BT}{2\pi\hbar^2}\right)^{\frac{3}{2}}}e^{-\beta E}dE$$

$$= \frac{2}{\pi^2}\left(\frac{\pi}{k_BT}\right)^{\frac{3}{2}}E^{\frac{1}{2}}e^{-\beta E}dE. \tag{5.7}$$

We can move from a probability density in energy to a probability density in velocity (or equivalently momentum $\mathbf{p} = m\mathbf{v}$) using $E = \frac{1}{2}mv^2$, where the speed $v$ is the magnitude of the velocity $\mathbf{v}$, which implies $dE = mv\,dv$. This gives

$$p(v)dv = \left(\frac{m}{2\pi k_BT}\right)^{\frac{3}{2}}e^{-\frac{mv^2}{2k_BT}}4\pi v^2 dv. \tag{5.8}$$

The probability density we have found in Eq. (5.8) is implicitly one in phase space, $P(\mathbf{x},\mathbf{p})$, but since the energy of a gas molecule is independent of its position and its direction of motion, we can write

$$\int d^3\mathbf{x}\int d^3\mathbf{p}\,P(\mathbf{x},\mathbf{p}) = \int_0^\infty dv\,p(v) = 1. \tag{5.9}$$

From Eq. (5.9), if we treat all the particles in the gas as equivalent, we can infer that the number density of particles that have velocities within $d^3\mathbf{v} = dv_x\,dv_y\,dv_z$ of $\mathbf{v}$, which we will denote by $f(\mathbf{v})d^3\mathbf{v}$, satisfies

$$n_0 = \frac{N}{V} = \frac{N}{V}\int d^3\mathbf{x}\int d^3\mathbf{p}\,P(\mathbf{x},\mathbf{p}) = n_0\int_0^\infty dv\,p(v) = \int d^3\mathbf{v}f(\mathbf{v}), \tag{5.10}$$

where $n_0$ is the number density of the particles. By rewriting the integral on the right-hand side and noting that there is no preferred direction in space (so $f(\mathbf{v})$ should be a function of $v$ only), we see that

$$n_0\int_0^\infty dv\,p(v) = \int_0^{2\pi}d\phi\int_0^\pi d\theta\,\sin\theta\int_0^\infty v^2 dv f(\mathbf{v}) = \int_0^\infty 4\pi v^2 dv f(\mathbf{v}), \tag{5.11}$$

which allows us to read off the **Maxwell–Boltzmann velocity distribution** for particle velocities in an ideal gas:

$$f(\mathbf{v}) = n_0\left(\frac{m}{2\pi k_BT}\right)^{\frac{3}{2}}e^{-\frac{mv^2}{2k_BT}}. \tag{5.12}$$

To obtain this result, we used the fact that in an ideal gas there are no interactions between particles, so the distribution of values of velocity for a single particle will mirror the distribution of values of velocity for a gas with many particles.

## 5.2 Properties of the Maxwell–Boltzmann Velocity Distribution

We now identify some important features of the Maxwell–Boltzmann velocity distribution. The first is the normalization such that

$$\int d^3\mathbf{v} f(\mathbf{v}) = n_0, \tag{5.13}$$

where $n_0$ is the number density of the gas. In our derivation we assumed that there is no flow, so that we are in the rest frame of the fluid and $\langle \mathbf{v} \rangle = 0$.[1] If there is a net flow, so that $\langle \mathbf{v} \rangle \neq 0$, then we can shift from the rest frame of the fluid, in which case

$$f(\mathbf{v}) = n_0 \left( \frac{m}{2\pi k_B T} \right)^{\frac{3}{2}} e^{-\frac{m(\mathbf{v}-\langle \mathbf{v} \rangle)^2}{2k_B T}}. \tag{5.14}$$

We set $\langle \mathbf{v} \rangle = 0$ in our following discussion.

Recalling that $v^2 = v_x^2 + v_y^2 + v_z^2$, we see that the Maxwell–Boltzmann distribution factorizes into three identical distributions for $v_x, v_y, v_z$, i.e.

$$f(\mathbf{v}) = n_0 g(v_x) g(v_y) g(v_z), \tag{5.15}$$

where

$$g(v_x)dv_x = \frac{1}{n_0} dv_x \int \int dv_y dv_z f(\mathbf{v}) = \left( \frac{m}{2\pi k_B T} \right)^{\frac{1}{2}} e^{-\frac{m v_x^2}{2 k_B T}} dv_x, \tag{5.16}$$

and $n_0 g(v_x) dv_x$ is the density of particles with velocities in the $x$ direction between $v_x$ and $v_x + dv_x$. We see that $g(v_x)$ is a Gaussian distribution peaked at $v_x = 0$; $g(v_y)$ and $g(v_z)$ have similar forms. This observation can be helpful for calculating moments of the velocity distribution

It is helpful to calculate the first two moments of the distribution, which correspond to the average speed and mean square velocity, respectively (when $\langle \mathbf{v} \rangle = 0$). The first moment is

$$\langle v \rangle = \frac{\int d^3\mathbf{v}\, v\, f(\mathbf{v})}{\int d^3\mathbf{v} f(\mathbf{v})}$$

$$= \frac{1}{n_0} n_0 \left( \frac{m}{2\pi k_B T} \right)^{\frac{3}{2}} 4\pi \int_0^\infty dv\, v^3\, e^{-\frac{m v^2}{2 k_B T}}$$

$$= \left( \frac{m}{2\pi k_B T} \right)^{\frac{3}{2}} 4\pi \left( \frac{2 k_B T}{m} \right)^2 \int_0^\infty dx\, x^3\, e^{-x^2}$$

$$= \sqrt{\frac{8 k_B T}{\pi m}}$$

$$\simeq 1.60 \sqrt{\frac{k_B T}{m}}, \tag{5.17}$$

---

[1] A more sophisticated derivation allowing for flow follows from solving the Boltzmann equation allowing for collisions between gas molecules.

where we let $x^2 = mv^2/2k_BT$ in going from the second to the third line to obtain Eq. (5.17). The second moment can be calculated as

$$\langle v^2 \rangle = \frac{3k_BT}{m},\tag{5.18}$$

either by direct integration, or using the equipartition theorem. This corresponds to a root mean square velocity of

$$v_{\text{rms}} = \sqrt{\frac{3k_BT}{m}} \simeq 1.73\sqrt{\frac{k_BT}{m}}.\tag{5.19}$$

## 5.2.1 Distribution of Speeds

We can also determine a distribution of speeds (rather than velocities) by rewriting the Maxwell–Boltzmann velocity distribution, i.e.

$$\begin{aligned}F(v)dv &= \int_0^{2\pi} d\phi \int_0^{\pi} \sin\theta \, d\theta f(\mathbf{v}) \, v^2 \, dv \\ &= 4\pi v^2 f(\mathbf{v})dv \\ &= 4\pi n_0 \left(\frac{m}{2\pi k_BT}\right)^{\frac{3}{2}} v^2 e^{-\frac{mv^2}{2k_BT}} \, dv.\end{aligned}\tag{5.20}$$

A third velocity scale that enters naturally when we look at the distribution of speeds is the most probable speed $v_p$, which corresponds to the maximum in $F(v)$. This maximum is given by

$$\frac{dF}{dv} = 0,\tag{5.21}$$

which implies

$$\left(2v - \frac{mv^3}{k_BT}\right) e^{-\frac{mv^2}{2k_BT}} = 0\tag{5.22}$$

when $v = v_p$, i.e.

$$v_p = \sqrt{\frac{2k_BT}{m}} \simeq 1.41\sqrt{\frac{k_BT}{m}}.\tag{5.23}$$

We note that $v_p$, $\langle v \rangle$ and $v_{\text{rms}}$ are all very similar in magnitude, but are ordered as $v_p < \langle v \rangle < v_{\text{rms}}$. Which one is relevant in a given physical situation depends on the question being asked. These speeds and the entire distribution are illustrated in Fig. 5.1.

## Example: Speed of Molecules in Air

We can use the results we have obtained for $\langle v \rangle$, $v_{\text{rms}}$ and $v_p$ to calculate the speed of molecules in air. We approximate air as 78% nitrogen, 21% oxygen and 1% argon,[2] which

---

[2] The composition of dry air is more precisely given by 78.09% nitrogen, 20.95% oxygen, 0.93% argon and 0.04% carbon dioxide, with trace amounts of other gases. Water vapour is also present in the atmosphere; at sea level this is usually around the 1% level.

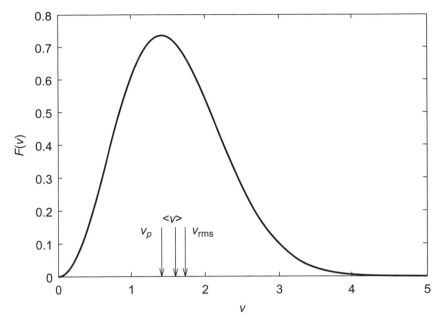

**Fig. 5.1**    Distribution of speeds that follows from the Maxwell–Boltzmann velocity distribution with positions of the average velocity $\langle v \rangle$, most probable velocity $v_p$ and root mean square velocity $v_{rms}$, marked. Velocity is measured in units of $\sqrt{\frac{k_B T}{m}}$ and the distribution is not normalized.

have molecular masses of 28.01 amu (nitrogen), 32.00 amu (oxygen) and 39.95 amu (argon). The average mass of a molecule in air is thus 29.0 amu, so we can calculate for air at a temperature of 300 K that

$$\sqrt{\frac{k_B T}{m}} \simeq \sqrt{\frac{1.3807 \times 10^{-23} \times 300}{29.0 \times 1.66 \times 10^{-27}}} = 293 \text{ m s}^{-1}. \tag{5.24}$$

Thus we get the following measures for the speed of molecules in air:

$$\langle v \rangle \simeq 1.60 \sqrt{\frac{k_B T}{m}} \sim 470 \text{ m s}^{-1},$$

$$v_{rms} \simeq 1.73 \sqrt{\frac{k_B T}{m}} \sim 510 \text{ m s}^{-1},$$

$$v_p \simeq 1.41 \sqrt{\frac{k_B T}{m}} \sim 410 \text{ m s}^{-1},$$

indicating that characteristic speeds of molecules in air are in the range 400–500 m s$^{-1}$.

## 5.3 Kinetic Theory for an Ideal Gas

We now make use of kinetic theory to calculate the properties of a dilute gas in a finite container.

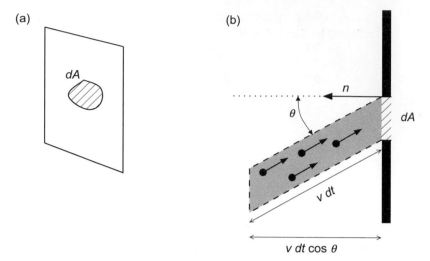

Fig. 5.2 (a) Schematic of area $dA$ on wall of container; (b) cylinder containing all molecules with a velocity directed at an angle $\theta$ to the wall with speed between $v$ and $v + dv$ that are available to collide with the element of area $dA$ in time $dt$.

## 5.3.1 Pressure in an Ideal Gas

Molecules in a gas collide with the walls of their container and exchange momentum with the walls, which leads to a force per unit area that can be identified as the pressure. We can calculate the pressure by determining the momentum transfer to the wall. To do this, we consider a fluid at rest and focus on an area $dA$ on the wall, as depicted in Fig. 5.2(a). If we consider a short time interval $dt$, we can calculate how many molecules are incident on $dA$ with velocities between $\mathbf{v}$ and $\mathbf{v} + d\mathbf{v}$. This will be exactly the number of molecules in the cylinder shown in Fig. 5.2(b), i.e.

$$\underbrace{f(\mathbf{v})\, d^3\mathbf{v}}_{\text{Density of molecules}} \times \underbrace{v_n\, dA\, dt}_{\text{Volume of cylinder}} \;,$$

where $v_n = \mathbf{v} \cdot \hat{\mathbf{n}}$ with $\hat{\mathbf{n}}$ the unit normal to the surface. We can also write $v_n = v \cos\theta$ in the co-ordinate system used in Fig. 5.2(b). To determine the total number of molecules colliding with $dA$, we need to integrate over all velocities with positive $v_n$ (molecules with negative $v_n$ move away from the surface).

Similarly, we can determine the momentum carried to the wall by the molecules in the cylinder as

$$\underbrace{(m\mathbf{v})\, f(\mathbf{v}) d^3\mathbf{v}}_{\text{Momentum density}} \times \underbrace{v_n\, dA\, dt}_{\text{Volume of cylinder}} \;,$$

hence the total incident momentum on the region $dA$ is

$$dA\, dt \int_{+} d^3\mathbf{v}\, m\mathbf{v}\, f(\mathbf{v})\, v_n,$$

where the + subscript on the integral sign means that $v_n > 0$, and the total momentum reflected by the surface is

$$dA\,dt \int_- d^3\mathbf{v}\, m\mathbf{v}\, f(\mathbf{v})\, (-v_n).$$

These two momenta point in opposite directions, so there is momentum $2mv_n$ transferred in each collision. The total momentum $d\mathbf{p}$ exchanged between the gas and the wall in time $dt$ via the surface $dA$ is thus

$$
\begin{aligned}
d\mathbf{p} &= dA\,dt \left[ \int_+ d^3\mathbf{v}\, m\mathbf{v}\, f(\mathbf{v})\, v_n - \int_- d^3\mathbf{v}\, m\mathbf{v}\, f(\mathbf{v})\, (-v_n) \right] \\
&= dA\,dt \int d^3\mathbf{v}\, m\mathbf{v}\, f(\mathbf{v})\, v_n \\
&= dA\,dt\, n_0\, m\, \langle v_n \mathbf{v} \rangle.
\end{aligned}
\tag{5.25}
$$

Identifying $d\mathbf{p}/dt$ as the force applied to the wall by molecular collisions, and dividing by $dA$, we have a force per unit area, which allows us to determine the force $\mathbf{F}$ per area $A$ on the surface with normal $\hat{\mathbf{n}}$ as

$$
\begin{aligned}
\frac{\mathbf{F}}{A} &= n_0\, m\, \langle v_n \mathbf{v} \rangle \\
&= \rho \sum_{i=1}^{3} \hat{n}_i \langle v_i \mathbf{v} \rangle,
\end{aligned}
\tag{5.26}
$$

where $\rho = n_0\, m$ is the mass density of the gas. We notice that in the expression Eq. (5.26), $\langle v_n \mathbf{v} \rangle = 0$ for any of the components of $\mathbf{v}$ that are orthogonal to $\hat{\mathbf{n}}$ for an ideal gas.[3]

The force per unit area exerted on the wall by the fluid is the stress, $\Sigma$, which we can write as (note the minus sign, which arises because the force is directed oppositely to the normal to the surface)

$$\Sigma_i = -\frac{F_i}{A} = \sum_{i=1}^{3} \sigma_{ij} \hat{n}_j, \tag{5.27}$$

where $\sigma_{ij}$ is the stress tensor. We can read off from Eqs (5.26) and (5.27) that[4]

$$\sigma_{ij} = -\rho \left\langle v_i v_j \right\rangle. \tag{5.28}$$

---

[3] To illustrate this point, suppose that $\hat{\mathbf{n}} = \hat{\mathbf{z}}$, in which case $v_n = v_z$, and then

$$\langle v_z \mathbf{v} \rangle = \begin{bmatrix} \langle v_z v_x \rangle \\ \langle v_z v_y \rangle \\ \langle v_z^2 \rangle \end{bmatrix} = \frac{1}{n_0} \int_{-\infty}^{\infty} dv_x \int_{-\infty}^{\infty} dv_y \int_{-\infty}^{\infty} dv_z \begin{bmatrix} v_z v_x \\ v_z v_y \\ v_z^2 \end{bmatrix} f(\mathbf{v}),$$

and since $f(\mathbf{v})$ is an even function of $v_x$, $v_y$ and $v_z$, we see that $\langle v_z v_x \rangle = \langle v_z v_y \rangle = 0$ and only $\langle v_z^2 \rangle \neq 0$.

[4] Recall that we assumed a fluid at rest. If a fluid is in motion with an average velocity of $\langle \mathbf{v} \rangle = \mathbf{u}$, then we can write the velocity of molecules in the fluid as $\mathbf{v} = \mathbf{u} + \mathbf{w}$, where $\mathbf{w}$ describes the random motion of molecules in the fluid and $\langle \mathbf{w} \rangle = 0$. The stress tensor is defined to only include the random motion of the molecules and not any overall flow, and hence in general takes the form

$$\sigma_{ij} = -\rho \left\langle w_i w_j \right\rangle.$$

Our discussion indicates that the stress tensor is (in principle) not necessarily isotropic and can be represented by the $3 \times 3$ matrix $\sigma_{ij}$, which is a second-rank tensor. The component $\sigma_{ij}$ is the force per unit area in the $j$ direction exerted on a surface with normal in the $i$ direction. The off-diagonal elements correspond to a shear stress (forces applied parallel to the surface) and the diagonal elements to a tensile stress (forces applied normal to the surface).

A gas cannot support shear stresses, so the off-diagonal components of the stress tensor vanish (which can also be checked by explicit calculation for the Maxwell–Boltzmann velocity distribution). Hence

$$\sigma_{ii} = -\rho \left\langle v_i^2 \right\rangle, \tag{5.29}$$

and $\left\langle v_x^2 \right\rangle = \left\langle v_y^2 \right\rangle = \left\langle v_z^2 \right\rangle = \frac{1}{3} \left\langle v^2 \right\rangle$ due to the isotropy of the Maxwell–Boltzmann velocity distribution, as can be verified by direct computation. Thus the stress is isotropic and has mean value

$$\sigma = -\frac{1}{3} n_0 m \left\langle v^2 \right\rangle = -\frac{1}{3} n_0 m \frac{3 k_B T}{m} = -n_0 k_B T = -\frac{N k_B T}{V}, \tag{5.30}$$

which corresponds to a force per unit area exerted on the surface (i.e. a pressure)

$$P = \frac{F}{A} = \frac{N k_B T}{V}, \tag{5.31}$$

as we expect for an ideal gas. Provided we consider the special case of isotropic fluids (i.e. fluids which have the same properties in all directions), such as an ideal gas, we only need to worry about the pressure and none of the other components of the stress tensor. However, when dealing with complex fluids which may be anisotropic, such as liquid crystals, the tensor nature of stress becomes important.

### 5.3.2 Effusion

In our calculation of the pressure, we calculated the number of particles striking the area $dA$ in time $dt$ for a given velocity $\mathbf{v}$. Integrating over all velocities (subject to the restriction that $v_n > 0$), the total number of particles striking the area $dA$ is

$$N_{\text{incident}} = dA \, dt \int_+ d^3 \mathbf{v} \, f(\mathbf{v}) \, v_n. \tag{5.32}$$

In effusion we consider the situation when the area $dA$ is removed, so that there is a small hole in the side of the container and gas molecules can escape from the container (Fig. 5.3). In that case, the value of $N_{\text{incident}}$ in Eq. (5.32) is the total number of particles escaping the container through the hole in time $dt$. Hence we can calculate the flux of molecules escaping through the hole per unit area per unit time as

$$\int_+ d^3 \mathbf{v} f(\mathbf{v}) v_n = \int_0^{2\pi} d\phi \int_0^{\frac{\pi}{2}} d\theta \sin\theta \int_0^\infty v^2 \, dv \, v \cos\theta f(\mathbf{v})$$

$$= 2\pi \int_0^1 dx \, x \int_0^\infty v^2 \, dv \, v f(\mathbf{v})$$

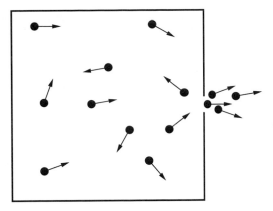

Fig. 5.3 Illustration of effusion – particles can escape from a small hole in a container.

$$= 2\pi \left[\frac{1}{2}x^2\right]_0^1 \int_0^\infty v^2 \, dv \, v f(\mathbf{v})$$

$$= \pi \int_0^\infty v^2 \, dv \, v f(\mathbf{v})$$

$$= \frac{n_0}{4} \langle v \rangle , \tag{5.33}$$

where we used the substitution $x = \cos\theta$ and recalled that the Maxwell–Boltzmann velocity distribution is isotropic in $\mathbf{v}$, and that the angular integral which appears in the expectation value $\langle v \rangle$ is

$$\int_0^{2\pi} d\phi \int_0^\pi d\theta \, \sin\theta = 4\pi. \tag{5.34}$$

One application of effusion is in the enrichment of uranium. The rate of effusion we calculated above is

$$\frac{n_0}{4} \langle v \rangle = \frac{1}{4} n_0 \sqrt{\frac{8k_B T}{\pi m}} \simeq 0.4 n_0 \sqrt{\frac{k_B T}{m}}, \tag{5.35}$$

which depends on the mass of the molecules. Uranium hexafluoride gas molecules will have slightly different masses depending on whether they contain the $^{235}$U isotope or the $^{238}$U isotope, and hence slightly different rates of effusion. By repeated application of effusion it is possible to increase the proportion of the isotope $^{235}$U which is used for fission in either nuclear reactors or nuclear weapons.

## 5.4 Brownian Motion

During Brownian motion a particle in a fluid moves in a seemingly random way, as was observed for pollen grains in water by Robert Brown. This motion comes from the interaction of the particle with the smaller molecules in the fluid. We will illustrate Brownian motion using a one-dimensional example for simplicity. Let the position of the particle of interest, e.g. a pollen grain, be $x$, in which case its velocity is given by $v = dx/dt$. The total force experienced by the particle will have two contributions: (i) an external force, $\mathcal{F}(t)$ and (ii) a net force, $F(t)$ that arises from the interactions of the particles with the molecules in the fluid.[5] Newton's second law gives

$$m\frac{dv}{dt} = \mathcal{F}(t) + F(t). \tag{5.36}$$

Now, $F(t)$ arises from the collisions of molecules in the fluid with the particle of interest. These molecules are moving randomly at very fast speeds and hence $F(t)$ varies rapidly in time in both sign and magnitude. If we are interested in the average motion of the particle, rather than for a very short period of time, then we can average over $F(t)$. Suppose that $\tau_0$ is the time scale over which $F(t)$ fluctuates, then for times $T \gg \tau_0$, there is no net force from fluctuations, i.e.

$$\langle F \rangle = \frac{1}{T} \int_0^T d\tilde{t}\, F(\tilde{t}) = 0. \tag{5.37}$$

To take advantage of this observation about the nature of $F(t)$, write the velocity as $v = u + \delta v$, where $\delta v$ fluctuates on time scales of order $\tau_0$ so that the time average $\langle \delta v \rangle = 0$ and hence $u$ gives the slower average motion on time scales $\gg \tau_0$.

We expect that the force on the particle from the fluid should act so as to bring the particle to equilibrium, so it will depend on the particle velocity, in which case we can write the fluctuating force in the form

$$F(t) = F_u(t) + \delta F(t), \tag{5.38}$$

where $\langle \delta F(t) \rangle = 0$ and $F_u(t)$ is a slowly varying part of the force, that we can expand in powers of $u$. When $u = 0$, we expect that $F_{u=0}(t) = 0$, and so the lowest term in the expansion will be the linear term. Hence we may approximate

$$F_u(t) \simeq -\alpha u, \tag{5.39}$$

where $\alpha$ is a constant that quantifies the damping of motion of the particle by the fluid. This gives us the following equation for the slowly varying part of the velocity:

$$m\frac{du}{dt} = \mathcal{F}(t) - \alpha u. \tag{5.40}$$

We can thus rewrite Eq. (5.36) to obtain the **Langevin equation** (Langevin, 1908)

$$m\frac{dv}{dt} = \mathcal{F}(t) - \alpha v + \delta F(t), \tag{5.41}$$

---

[5] The quantitative link between the idea that molecules in the fluid exert random forces on a particle was provided by Einstein in 1905, the same year that he also published his famous papers on special relativity and the photoelectric effect.

where we assumed that $\alpha u \simeq \alpha v$ since the error we have introduced, $\alpha \delta v$, should be negligible in comparison to $\delta F(t)$.

We first consider the Langevin equation in the absence of an external force $\mathcal{F}(t)$. It is instructive to consider the situation when there is no random force, $\delta F(t)$, in which case we have

$$m\frac{dv}{dt} = -\alpha v, \tag{5.42}$$

which has the solution

$$v(t) = v_0 e^{-\frac{t}{\tau}}, \tag{5.43}$$

where we defined $\tau = m/\alpha$. In this case, the velocity drops to zero exponentially and the solution of the equation does not accord with the phenomenology of Brownian motion.

We now reintroduce the rapidly fluctuating force term. Recall that $v = \dot{x} = dx/dt$ and multiply both sides of Eq. (5.41) by $x$, in which case we get

$$mx\frac{d\dot{x}}{dt} = m\frac{d(x\dot{x})}{dt} - m\dot{x}^2 = -\alpha x\dot{x} + x\delta F(t). \tag{5.44}$$

Now, if we take an average over times $\gg \tau_0$, this becomes

$$m\left(\frac{d}{dt} + \alpha\right)\langle x\dot{x}\rangle = m\left\langle \dot{x}^2\right\rangle + \langle x\delta F(t)\rangle, \tag{5.45}$$

and since the fluctuating force is uncorrelated with position, and also has vanishing mean (i.e. $\langle \delta F(t)\rangle = 0$), we have $\langle x\delta F(t)\rangle = \langle x\rangle\langle \delta F(t)\rangle = 0$. We can also use the equipartition theorem to evaluate $m\left\langle \dot{x}^2\right\rangle$, which is twice the kinetic energy, and hence $m\left\langle \dot{x}^2\right\rangle = k_B T$.[6]

Thus we can simplify Eq. (5.45) to

$$\left(\frac{d}{dt} + \frac{\alpha}{m}\right)\langle x\dot{x}\rangle = \frac{k_B T}{m}, \tag{5.46}$$

which has the solution (as can be verified by substitution)

$$\langle x\dot{x}\rangle = \frac{k_B T}{\alpha}\left(1 - e^{-\frac{t}{\tau}}\right), \tag{5.47}$$

where $\tau = m/\alpha$ takes the same value as in the absence of a random force and is the characteristic time scale for velocity relaxation for the particle. If we note that

$$\frac{d}{dt}\left\langle x^2(t)\right\rangle = 2\langle x\dot{x}\rangle$$

$$= \frac{2k_B T}{\alpha}\left(1 - e^{-\frac{t}{\tau}}\right), \tag{5.48}$$

then we can integrate to get

$$\left\langle x^2(t)\right\rangle = \frac{2k_B T}{\alpha}\left[t + \tau e^{-\frac{t}{\tau}} + C\right], \tag{5.49}$$

---

[6] It might seem that we reintroduced the fluctuating force term only for it to drop out of the equation again when we took a temporal average. However, the presence of fluctuating forces allows for equilibrium to be established, which then allows us to use the equipartition theorem.

where $C$ is a constant of integration. Requiring that $\langle x^2(t=0)\rangle = 0$ implies that $C = -\tau$, and hence

$$\langle x^2(t)\rangle = \frac{2k_BT}{\alpha}\left[t - \tau\left(1 - e^{-\frac{t}{\tau}}\right)\right]. \tag{5.50}$$

We can explore the behaviour of this expression in both the short and long time limits. For very short times, $t \ll \tau$, we can Taylor expand the exponential in Eq. (5.50) as

$$e^{-\frac{t}{\tau}} \simeq 1 - \frac{t}{\tau} + \frac{1}{2}\frac{t^2}{\tau^2} + \cdots, \tag{5.51}$$

and obtain

$$\langle x^2(t)\rangle \simeq \frac{k_BT}{\alpha\tau}t^2, \tag{5.52}$$

while in the long time limit $t \gg \tau$, we obtain

$$\langle x^2(t)\rangle \simeq \frac{2k_BT}{\alpha}t. \tag{5.53}$$

The solutions to the Langevin equation we found in Eqs (5.52) and (5.53) mirror the behaviour we found for the random walk in Section 1.2.4, that at short times the rms displacement grows ballistically whereas for longer times it grows diffusively. If we record the position of a Brownian particle then its displacement will have the form of a random walk, as illustrated for one dimension in Fig. 1.4.

## 5.5  Diffusion

In our discussion of random walks in Chapter 1 (and in Problem 1.4) we saw that the continuum limit of a random walk corresponds to diffusion, in which a probability distribution in space and time, $\rho(\mathbf{x}, t)$, evolves according to the diffusion equation

$$\frac{\partial\rho}{\partial t} = D\nabla^2\rho, \tag{5.54}$$

where $D$ is the diffusion coefficient. The form of diffusion equation in Eq. (5.54), with one time and two space derivatives, hints at the relationship between characteristic time ($t_c$) and length ($L_c$) scales in diffusion processes, that is $L_c \sim \sqrt{t_c}$.

### 5.5.1  Mean Free Path and Collision Time

We can use kinetic theory to get insight into the microscopic origin of the diffusion coefficient $D$. To do this we will consider collisions, which correspond to when particles get close enough to each other to interact strongly. Rather than trying to describe every collision precisely, we treat them statistically, and characterize collisions with a few parameters. The first of these is the mean distance between collisions, the mean free path, which we denote by $l$.

In order to visualize the mean free path, treat the molecules in a gas as spheres of diameter $d$. There will be a collision if the centres of two spheres pass within a distance $d$

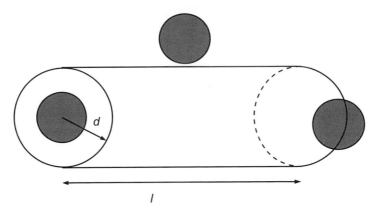

Fig. 5.4 Cylinder swept out between collisions for spherical molecules of diameter $d$ in a gas with mean free path $l$.

of each other. In between collisions, there are no forces acting on the molecules, so they will travel in straight lines and sweep out a cylindrical volume of $l\pi d^2$, as illustrated in Fig. 5.4.[7]

On average, there should be a collision every time this cylindrical volume is equal to the volume per atom, $1/n$, where $n$ is the number density, so

$$l\pi d^2 = \frac{1}{n}, \tag{5.55}$$

and so the mean free path is

$$l = \frac{1}{n\pi d^2}. \tag{5.56}$$

Having obtained a distance scale for collisions, we can also determine a time scale. For molecules moving with speed $v$, the collision time $\tau$ will be related to the mean free path by

$$l = v\tau. \tag{5.57}$$

If we take $v$ to be the rms speed $v_{\mathrm{rms}}$, introduced in Eq. (5.19), and use the expression we obtained above for the mean free path, Eq. (5.56), then we get an estimate[8]

$$\tau \simeq \frac{l}{v_{\mathrm{rms}}} = \sqrt{\frac{m}{3\pi^2 k_B T}}\frac{1}{nd^2}. \tag{5.58}$$

Hence we see that the collision time scales as $\tau \sim T^{-\frac{1}{2}}$. This is in accord with our intuition that at higher temperatures molecules will move more quickly and have

[7] Strictly, if we imagine the molecules interacting via a potential similar to the Lennard–Jones potential introduced in Eq. (4.111), the collisions correspond to when molecules are close enough to experience the strongly repulsive part of the potential, and in between collisions there are very weak attractions between the molecules. To a first approximation we can ignore the weak attractions that will cause molecules to deviate from straight-line motion and assume that only the repulsive forces are important, in which case the molecules move in straight lines between collisions.

[8] Our first instinct might be that we should take $v$ to be the average speed $\langle v \rangle$, but because the molecules are moving relative to each other, we should have the average relative speed, which a careful analysis reveals to be $v_{\mathrm{rms}}$.

more collisions per unit time (and hence less time between collisions) than at low temperatures.

## Example: Mean Free Path and Collision Time in Air

We can use Eqs (5.56) and (5.58) to estimate the mean free path and collision time for molecules in air. To calculate the mean free path, we need the number density $n$ and the diameter $d$ for molecules in air. We can obtain the number density at atmospheric pressure $(1.013 \times 10^5$ Pa$)$ and a temperature of 300 K from the ideal gas law as

$$n = \frac{P}{k_B T} = \frac{1.013 \times 10^5}{1.3807 \times 10^{-23} \times 300} \simeq 2.4 \times 10^{25} \text{ m}^{-3}.$$

We can also note that the average bond length in $N_2$ is $\sim 1.1$ Å and the average bond length in $O_2$ is $\sim 1.2$ Å, so if we treat air to be 80% nitrogen and 20% oxygen, this gives $d \simeq 1.1$ Å. Hence the mean free path in air is

$$l = \frac{1}{n\pi d^2} = \frac{1}{2.4 \times 10^{25} \times \pi \times (1.1 \times 10^{-10})^2} \simeq 1\ \mu\text{m}.$$

Once we have the mean free path we can use our earlier estimate for the speed of molecules in air of $v_{\text{rms}} = 510 \text{ m s}^{-1}$ to get a collision time of

$$\tau \simeq \frac{l}{v_{\text{rms}}} \simeq \frac{10^{-6}}{510} \simeq 2 \text{ ns}.$$

## 5.5.2 Fick's Law

Consider a situation in which the average density of molecules is constant, but that some subset of the molecules is labelled, e.g. by being in a long-lived excited nuclear state, and this subset of molecules is not uniformly distributed, so that their number density, $n$, is position dependent. Further, simplify to the case in which the non-uniformity is only in the $z$ direction. Collisions between the labelled and unlabelled molecules will tend to mix the two components so that after a sufficiently long time the labelled molecules are distributed uniformly in space. We can investigate the relaxation from the original out-of-equilibrium initial condition towards a uniform equilibrium state via diffusion.

If we consider a plane at constant $z$ we can investigate the flux of labelled particles across that plane, $J_z$, which will be the mean number of labelled particles crossing the plane per unit area per unit time. We can apply the ideas of kinetic theory that we have developed in this chapter to calculate this flux in a gas by using a similar approach to Fig. 5.2, as illustrated in Fig. 5.5. If we let the number density of particles at a given $z$ be $n(z)$, then we can calculate the flux coming up through the plane at fixed $z$ for a given $\theta$ as

$$J_z = \frac{dN}{dAdt} = v_z \times \underbrace{\frac{n(z - l\cos\theta)}{n_0}}_{\text{Number of molecules}} f(\mathbf{v})d^3\mathbf{v}, \tag{5.59}$$

where we have assumed that particles travel a mean free path before crossing the plane (the factor of $\cos\theta$ is because the particle velocities are not guaranteed to be normal to the

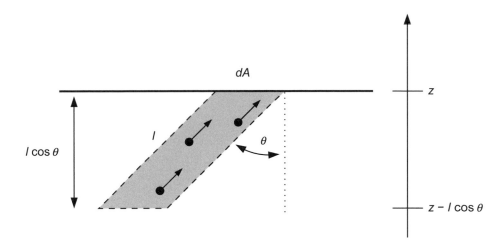

Fig. 5.5 Geometry used for obtaining Fick's law. Particles are assumed to travel a mean free path $l$ before crossing the plane at $z$ with an incoming angle of $\theta$.

surface and hence the change in $z$ for a particle travelling a mean free path and having a velocity oriented at angle $\theta$ to the plane is $l\cos\theta$). In practice, not all molecules will have travelled as far as a mean free path before reaching our imaginary surface and so, while our estimate of the diffusion coefficient should capture the correct qualitative behaviour, a more sophisticated treatment would give a more quantitatively accurate result. We can perform a similar estimate to Eq. (5.59) for the flux coming from above the plane and when we sum the two contributions we get Fick's law:

$$
\begin{aligned}
J_z &= \frac{1}{n_0}\int d^3\mathbf{v}\, v\cos\theta\,[n(z-l\cos\theta)-n(z+l\cos\theta)]\,f(\mathbf{v}) \\
&\simeq -\frac{2}{n_0}\left\{\int d^3\mathbf{v}\, vl\cos^2\theta f(\mathbf{v})\right\}\frac{\partial n}{\partial z} \\
&= -D\frac{\partial n}{\partial z},
\end{aligned}
\tag{5.60}
$$

where we expanded

$$
n(z\pm l\cos\theta)\simeq n(z)\pm l\cos\theta\frac{\partial n}{\partial z},
\tag{5.61}
$$

and combined all of the terms in front of the gradient into a constant:

$$
\begin{aligned}
D &= 2l\frac{\int_0^{\frac{\pi}{2}} d\theta\,\sin\theta\cos^2\theta}{\int_0^{\pi} d\theta\,\sin\theta}\,\langle v\rangle \\
&= 2l\frac{\left[-\frac{1}{3}\cos^3\theta\right]_0^{\frac{\pi}{2}}}{[-\cos\theta]_0^{\pi}}\,\langle v\rangle \\
&= \frac{1}{3}l\,\langle v\rangle.
\end{aligned}
\tag{5.62}
$$

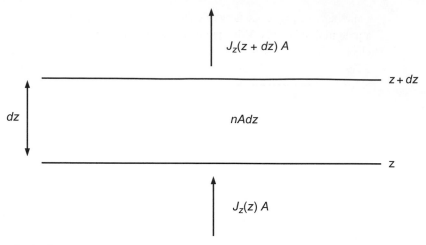

**Fig. 5.6** Number of molecules in a slab of thickness $dz$ and area $A$, $n(z)Adz$, and the flux of molecules entering $AJ_z(z)$ and leaving $AJ_z(z + dz)$ the slab.

The form of Fick's law, Eq. (5.60), implies that the concentration $n$ satisfies the diffusion equation. To see this, consider a thin slab of thickness $dz$ with lower surface at $z$ and upper surface at $z + dz$, as illustrated in Fig. 5.6. The rate of change of the number of labelled particles in the slab, $nAdz$, will be given by the difference in particle numbers entering through the two surfaces per unit time, where we note that the rate at which particles enter or leave the slab will be given by the area times the flux, i.e.

$$\frac{\partial}{\partial t}(nAdz) = AJ_z(z) - AJ_z(z + dz), \tag{5.63}$$

which we can simplify to

$$\frac{\partial n}{\partial t}dz = -\frac{\partial J_z}{\partial z}dz, \tag{5.64}$$

and when we subsitute in Eq. (5.60) for $J_z$, we obtain the diffusion equation for $n$:

$$\frac{\partial n}{\partial t} = D\frac{\partial^2 n}{\partial z^2}, \tag{5.65}$$

confirming that $D$ is the diffusion coefficient.

# 5.6 Transport

Kinetic theory is particularly useful for calculating properties of systems that are perturbed so that there is a flow of some quantity across the system, which is known as transport. The example of diffusion in Section 5.5 is an example of transport: the particle flux $J_z$ is related to the gradient of the particle density gradient by the diffusion coefficient $D$. Other examples of transport include the flow of energy, entropy or charge.

In the example of the flow of charge under the influence of an electric field, suppose that a potential difference $V$ is applied across a distance $L$ in a metal, giving rise to a uniform electric field $E = V/L$. Then the potential difference gives rise to a flow of charge, i.e. a current I. Provided the electric field is not too strong, the current will be related to the potential difference by

$$I = GV, \tag{5.66}$$

where $G$ is the conductance. The inverse of the conductance is the resistance $R = 1/G$, so that Eq. (5.66) is equivalent to Ohm's law:

$$V = IR. \tag{5.67}$$

It is often more convenient to work with current density, $\mathbf{j} = I/A$, the current per unit area, with $A$ the cross-sectional area of the metal, in which case we can write the relation as

$$j = \sigma E, \tag{5.68}$$

where $\sigma = G/AL$ is the conductivity, which has the advantage that it does not depend on the dimensions of the sample. This also has a similar form to Fick's law, Eq. (5.60), in that a current density ($j$) is proportional to a gradient ($E = -dV/dx$).

## 5.7 Summary

Kinetic theory gives a framework to calculate macroscopic physical properties of systems from the motion of their microscopic degrees of freedom. We can apply this to an ideal gas to find that the distribution of particle velocities in an ideal gas is given by the Maxwell–Boltzmann velocity distribution

$$f(\mathbf{v}) = n_0 \left( \frac{m}{2\pi k_B T} \right)^{\frac{3}{2}} e^{-\frac{mv^2}{2k_B T}}. \tag{5.69}$$

The Maxwell–Boltzmann velocity distribution can be shown to reproduce the ideal gas law, and used to calculate the rate of effusion.

When collisions between molecules are important, diffusion can take place. For a small object that is still much larger than molecular length scales, such as a pollen grain, subjected to molecular collisions, Brownian motion can occur, which is described by the Langevin equation

$$m\frac{dv}{dt} = \mathcal{F}(t) - \alpha v + \delta F(t). \tag{5.70}$$

At the molecular scale, important parameters for describing collisions between molecules are the mean free path $l$ and the collision time $\tau$.

# Problems

**5.1**  Calculate the following quantities for a three-dimensional classical ideal gas:

(a) $\langle v_x \rangle$,

(b) $\langle v_x^2 \rangle$,

(c) $\langle v^2 v_x \rangle$,

(d) $\langle v_x^3 v_y \rangle$,

(e) $\langle \left( v_x + b v_y \right)^2 \rangle$ ($b$ a constant),

(f) $\langle v^2 v_x^2 \rangle$.

**5.2**  To cool atoms down to temperatures low enough to observe Bose–Einstein condensation (discussed in Chapter 9), experimentalists use the technique of evaporative cooling. The idea of the technique is to remove the most energetic particles, and then let the resulting distribution thermalize to a lower temperature than previously. We can illustrate this idea with a Maxwell–Boltzmann distribution. If we wish to remove the most energetic fraction $\alpha$ of the particles, then we can remove all particles with speed greater than some $\tilde{v}$. Show that

$$\alpha = \frac{4}{\pi^{\frac{1}{2}}} \int_{\tilde{x}}^{\infty} dx\, x^2 e^{-x^2},$$

and relate $\tilde{x}$ to $\tilde{v}$.

If the gas is allowed to thermalize after the most energetic particles are removed, then show that the relationship between the new and old temperatures, $T_{\text{new}}$ and $T_{\text{old}}$, is given by

$$T_{\text{new}} = \frac{8 T_{\text{old}}}{3\pi^{\frac{1}{2}}} \int_{0}^{\tilde{x}} dx\, x^4 e^{-x^2} < T_{\text{old}}.$$

Find the value of $\tilde{x}$ that gives $\alpha = 0.1$ and hence find $T_{\text{new}}$ in terms of $T_{\text{old}}$ (you will need to evaluate the integrals numerically).

**5.3**  Naturally occuring uranium primarily consists of the two isotopes $^{235}$U (0.7% abundance) and $^{238}$U (99.3% abundance). Only the lighter of the two isotopes is useful for nuclear fission. One method to increase the abundance of the lighter isotope is to use chemical processes to obtain the gas uranium hexafluoride $UF_6$ from natural uranium. This gas, consisting of 0.7% $^{235}UF_6$ and 99.3% $^{238}UF_6$, is then put into a container with a small hole, through which the gas can escape into vacuum, i.e. effusion.

(a) Explain why the fraction of the gas escaping from the container is richer in $^{235}UF_6$ than the mixture left in the container.

(b) Give an estimate of how much the abundance of $^{235}UF_6$ among the molecules escaping the container is increased compared to the initial abundance of 0.7%.

*Note:* the atomic mass of fluorine is 18.998 amu.

**5.4** The pressure at the top of Mount Everest is about one-third of atmospheric pressure (101.3 kPa) and the temperature can be on the order of $-30°C$. Calculate the mean free path in air at this location and compare it to the result we found in Section 5.5.1.

**5.5** Giant molecular clouds are found in interstellar space and are composed mainly of hydrogen ($H_2$) molecules. They are a location where star formation can take place. A typical giant molecular cloud has a mass of $\sim 2 \times 10^5 M_\odot$, where $M_\odot \simeq 2 \times 10^{30}$ kg is the mass of the Sun, a size of $\sim 150$ light years and an average temperature of 10 K. What is the mean free path and collision time scale for hydrogen molecules in such a cloud? (The H–H bond length is 0.74 Å.)

**5.6** Consider a fluid confined between two parallel plates, one with velocity $u_x = u_0$ and the other at rest (Fig. 5.7).

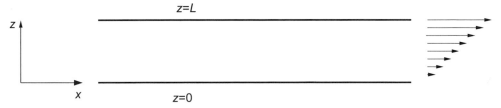

**Fig. 5.7** Velocity gradient for fluid confined between two parallel plates.

The motion of the plates sets up a velocity gradient within the fluid as the fluid next to each plate takes the velocity of the respective plate. Layers with different velocities exert a force on each other, because there is a change in momentum for particles that move between layers with different velocities. Writing this as a force per unit area, which acts in the $x$ direction and is exerted on a plane which is perpendicular to the $z$ direction, the force is proportional to the velocity gradient and is the coefficient of viscosity, $\eta$:

$$\sigma_{zx} = \eta \frac{\partial u_x}{\partial z}.$$

We will use kinetic theory to calculate the coefficient of viscosity.

The number of molecules in a fluid crossing a surface of unit area located in the plane $z = z_0$, and travelling in the $+z$ direction, is

$$\int_+ d\mathbf{v}\, v_z f(\mathbf{v}),$$

while the number of molecules crossing the surface in the opposite direction is

$$-\int_- d\mathbf{v}\, v_z f(\mathbf{v}).$$

(a) Show that if the distribution of velocities in the gas is given by a Maxwell–Boltzmann distribution, then both of these expressions are equal to

$$\frac{n_0 \langle v \rangle}{4}.$$

(b) Now consider the momentum carried between different layers of the fluid by the two sets of molecules (recalling that there is a velocity gradient in the $z$ direction). Assume that the molecules crossing the surface in the $+z$ direction originate on average from $z = z_0 - l$, where $l$ is the mean free path (the average distance between collisions of molecules). Hence show that the momentum transfer through the surface due to these molecules is

$$\frac{1}{4}\rho \langle v \rangle u_x|_{z=z_0-l}.$$

(c) Give the expressions for the momentum transferred by molecules travelling in the $-z$ direction and the net momentum transfer through the surface in the $+z$ direction.

(d) The net rate of momentum transfer found in parts (b) and (c) is equivalent to a force per unit area exerted in the $x$ direction by the gas below $z = z_0$ upon the gas above this surface, which is given by $\sigma_{zx}$. Assuming that $l$ is small on the scale of the experiment, find an expression for $\eta$.

**5.7**  You are on one side of a room in which the air is still, about 5 m from the far wall when hydrogen sulphide, $H_2S$ ("rotten egg gas") starts to leak in through a hole in the far wall. Estimate the time until you start to smell the poisonous gas.

**5.8**  Spectral lines from a hot body of gas may be broadened through the Doppler effect. For a spectral line which has wavelength $\lambda_0$ for molecules at rest, show that the combination of the motion of molecules in the gas and the Doppler effect leads to an intensity profile at wavelengths $\lambda$ near $\lambda_0$ of

$$I(\lambda) \propto \exp\left[-\frac{mc^2(\lambda - \lambda_0)^2}{2\lambda_0^2 k_B T}\right].$$

This is known as thermal Doppler broadening of a spectral line.

# 6 The Grand Canonical Ensemble

The restriction to fixed particle number that applies in the microcanonical and canonical ensembles is not always appropriate, e.g. for atoms adsorbing on a surface, or in chemical reactions, where the number of particles of different species can change. This necessitates the introduction of the grand canonical ensemble, where we consider a system that is in contact with both a thermal bath and a particle bath, so that the system can exchange both heat and particles with the respective baths. In this case the particle number $N = \sum_s n_s$ is not fixed, and instead only the mean value of $N$ is well defined. This is similar to the situation in which we go from having fixed energy $U$ in the microcanonical ensemble to a well-defined mean value of $U$ in the canonical ensemble. We start with a discussion of the chemical potential, a central concept in the grand canonical ensemble. It plays a similar role for a particle bath to the temperature of a heat bath. The utility of the grand canonical ensemble and the important role of the chemical potential (especially at low temperatures) will become particularly evident when we discuss the statistical mechanics of fermions and bosons in Chapters 7, 8 and 9.

## 6.1 Chemical Potential

Consider a system in thermal and diffusive contact with a much larger reservoir (so that both heat and particles may flow between the two). The total energy of the system + reservoir is $U_0$ and the total number of particles is $N_0$, both of which are fixed. The energy of the system is $E \ll U_0$ and the number of particles in the system is $N \ll N_0$, as illustrated in Fig. 6.1.

The total Helmholtz free energy, including both the bath and the system of interest, is

$$F_0 = F_R + F_S, \tag{6.1}$$

where $F_R$ is the reservoir (bath) free energy and $F_S$ is the Helmholtz free energy of the system. Now, $F_R$ is a function of $N_R = N_0 - N$ (the number of particles in the reservoir) and $F_S$ is a function of $N$, and we also know that at equilibrium with fixed $N_0$, $T$ and $V$, the Helmholtz free energy is minimized, so $dF_0 = 0$. Thus, if we hold the temperature $T$ and the volume $V$ constant:

$$dF_0 = \left.\frac{\partial F_R}{\partial N_R}\right|_{T,V} dN_R + \left.\frac{\partial F_S}{\partial N}\right|_{T,V} dN = 0, \tag{6.2}$$

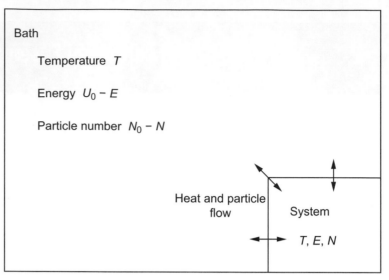

Fig. 6.1 System in contact with a heat and particle bath.

and recall that $N = N_0 - N_R$, which implies $dN = -dN_R$. We then obtain

$$\left( \frac{\partial F_R}{\partial N_R}\bigg|_{T,V} - \frac{\partial F_S}{\partial N}\bigg|_{T,V} \right) dN_R = 0, \tag{6.3}$$

which leads to the result that at equilibrium:

$$\frac{\partial F_R}{\partial N_R}\bigg|_{T,V} = \frac{\partial F_S}{\partial N}\bigg|_{T,V} = \mu, \tag{6.4}$$

where $\mu(T, V, N)$ is the chemical potential and is equal in two systems that are in equilibrium (this is similar to the argument we used to show that two systems in equilibrium have the same temperature). We can view the expression above as a definition of the chemical potential.

Now, $N$ can only take integer values, so the derivative should strictly be replaced by

$$\mu = F(T, V, N + 1) - F(T, V, N). \tag{6.5}$$

In the large-$N$ limit, we can consider the free energy as a continuous function and the distinction is generally unimportant. However, the discrete expression makes it clear that the chemical potential $\mu$ can be understood as the free energy required to add one particle to the system.

We can also relate the chemical potential to the entropy: consider the change in the entropy for an isothermal, isochoric process

$$dS = \frac{\partial S}{\partial U}\bigg|_{N,V} dU + \frac{\partial S}{\partial V}\bigg|_{U,N} dV + \frac{\partial S}{\partial N}\bigg|_{U,V} dN, \tag{6.6}$$

and since $dV = 0$, divide by $dN$ assuming fixed temperature $T$:

$$\frac{\partial S}{\partial N}\bigg|_{T,V} = \frac{1}{T}\frac{\partial U}{\partial N}\bigg|_{T,V} + \frac{\partial S}{\partial N}\bigg|_{U,V}, \tag{6.7}$$

therefore

$$T \frac{\partial S}{\partial N}\bigg|_{T,V} = \frac{\partial U}{\partial N}\bigg|_{T,V} + T \frac{\partial S}{\partial N}\bigg|_{U,V}. \tag{6.8}$$

If we take the derivative of $F = U - TS$ with respect to $N$ at constant $T$ and $V$, we also get

$$\mu = \frac{\partial F}{\partial N}\bigg|_{T,V} = \frac{\partial U}{\partial N}\bigg|_{T,V} - T \frac{\partial S}{\partial N}\bigg|_{T,V}, \tag{6.9}$$

hence, using Eq. (6.8) in Eq. (6.9) we get

$$\mu = -T \frac{\partial S}{\partial N}\bigg|_{U,V}. \tag{6.10}$$

We can thus rewrite Eq. (6.6) as

$$dS = \frac{1}{T}dU + \frac{P}{T}dV - \frac{\mu}{T}dN, \tag{6.11}$$

where we used the thermodynamic identity[1]

$$P = T \frac{\partial S}{\partial V}\bigg|_{U,N}. \tag{6.12}$$

We can also rearrange Eq. (6.11) to get the thermodynamic identity

$$dU = TdS - PdV + \mu dN. \tag{6.13}$$

## 6.1.1 Example: Ideal Gas

From the expression for the Helmholtz free energy of an ideal gas that we determined in Eq. (4.88) using the canonical ensemble:

$$F_{\text{monatomic}} = -N k_B T \ln \left( \frac{g_s n_Q}{n} \right) - N k_B T, \tag{6.14}$$

we can apply $\mu = \partial F / \partial N$ to determine that for a monatomic gas the chemical potential is

$$\mu = k_B T \ln \left( \frac{n}{g_s n_Q} \right), \tag{6.15}$$

where the result depends on the ratio of the number density, $n$, to the quantum concentration, $n_Q$ (introduced in Eq. (4.70)). For a gas with internal degrees of freedom, as we considered in Section 4.5, we need to consider additional contributions to the Helmholtz

---

[1] To derive the identity Eq. (6.12), recall that for fixed $N$

$$dS(U,V) = \frac{\partial S}{\partial U}\bigg|_{V,N} dU + \frac{\partial S}{\partial V}\bigg|_{U,N} dV,$$

and since the entropy is maximal at equilibrium, $dS = 0$, which in combination with $P = -\frac{\partial U}{\partial V}$ and $\frac{1}{T} = \frac{\partial S}{\partial U}$ gives the required result via

$$P = -\frac{\partial U}{\partial V}\bigg|_{S,N} = \frac{\frac{\partial S}{\partial V}\big|_{U,N}}{\frac{\partial S}{\partial U}\big|_{VN}} = T \frac{\partial S}{\partial V}\bigg|_{U,N}.$$

free energy. For example, for a diatomic gas at temperatures $T > \theta_r$, where rotations are important, but for which vibrations and electronic excitations can be ignored,

$$F = F_{\text{monatomic}} - Nk_BT \ln Z_r, \tag{6.16}$$

and hence

$$\mu = k_BT \ln\left(\frac{n}{g_s n_Q}\right) - k_BT \ln\left(Z_r\right), \tag{6.17}$$

recalling that $Z_r$ is independent of $N$ and takes the form $Z_r \simeq T/\theta_r$ at high temperatures. The schematic behaviour of the chemical potential for an ideal gas as a function of $n/n_Q$ is illustrated in Fig. 6.2. We will see, however, in Chapters 8 and 9 that the behaviour of $\mu$ in the quantum regime, $n/n_Q \gg 1$, differs from the predictions for a classical non-interacting gas.

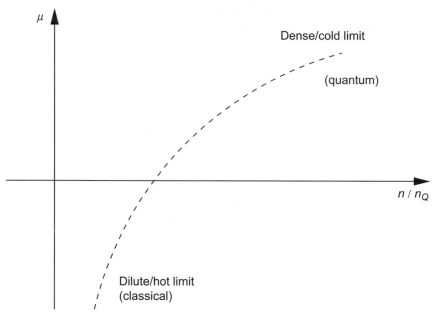

Fig. 6.2    Schematic plot of the chemical potential for an ideal gas as a function of $n/n_Q$.

We can use the ideal gas law to rewrite Eq. (6.17) as

$$\mu = k_BT \ln\left(\frac{n}{g_s n_Q Z_r}\right) = k_BT \ln\left(\frac{P}{g_s k_BT n_Q Z_r}\right), \tag{6.18}$$

which illustrates that chemical potential increases as pressure (equivalently density) increases. This matches our intuition that particles flow from regions of high concentration to low concentration, or from high pressure to low pressure. We know that for two systems in equilibrium, $\mu_1 = \mu_2$, so that if $\mu_1 > \mu_2$ and both systems contain ideal gases, then particles will flow from the higher chemical potential to the lower chemical potential system, leading $\mu_1$ to decrease and $\mu_2$ to increase until equilibrium is reached. This is analogous to temperature in that for two systems with different temperatures, heat will flow from the hotter to the colder system until an equilibrium temperature is reached.

It is also useful to invert Eq. (6.18) to write the pressure in terms of the chemical potential as

$$P = g_s k_B T n_Q Z_r e^{\beta\mu} = g_s k_B T n_Q Z_r z, \tag{6.19}$$

where the factor

$$z = e^{\beta\mu} \tag{6.20}$$

is referred to as the fugacity. When there are multiple species contributing to the pressure, with a partial pressure $P_i$ for the $i^{th}$ species, then the fugacity for each species, $z_i$, will be related to the partial pressure, $P_i$, in the same way as in Eq. (6.19).

## 6.2 Grand Canonical Partition Function

In the situation illustrated in Fig. 6.1, the temperature in the bath and the system is equal and the energy of the system, $E$, fluctuates about its mean value $U$, whilst the particle number of the system, $N$, fluctuates about a mean value $\mathcal{N}$. The probabilities $p_{Nj}$ of microstates with a given particle number $N$ and energy $E_j(N)$ (which depends on the particle number $N$) are subject to the constraints

$$\sum_{N=0}^{\infty}\sum_{j} p_{Nj} = 1; \quad \sum_{N=0}^{\infty}\sum_{j} p_{Nj}E_j(N) = U; \quad \sum_{N=0}^{\infty}\sum_{j} p_{Nj}N = \mathcal{N}. \tag{6.21}$$

We can determine the $p_{Nj}$ in a similar manner to the probabilities in the canonical and microcanonical ensembles by maximizing the entropy subject to the given constraints. The average particle number constraint forces us to introduce an additional Lagrange multiplier $\gamma$, so we vary the $p_{Nj}$ so as to maximize

$$S = -\sum_{N=0}^{\infty}\sum_{j} p_{Nj}\ln p_{Nj} - \lambda\left(\sum_{N=0}^{\infty}\sum_{j} p_{Nj} - 1\right) - \beta\left(\sum_{N=0}^{\infty}\sum_{j} p_{Nj}E_j(N) - U\right)$$
$$- \gamma\left(\sum_{N=0}^{\infty}\sum_{j} p_{Nj}N - \mathcal{N}\right). \tag{6.22}$$

The calculation proceeds in a similar manner to the calculation for the canonical ensemble, with the additional term multiplied by $\gamma$ applying the constraint on the average particle number, to give

$$\frac{\partial S}{\partial p_{Nj}} = -\left(\ln p_{Nj} + 1\right) - \lambda - \beta E_j(N) - \gamma N = 0, \tag{6.23}$$

which, when we solve for the $p_{Nj}$ and apply the constraint that their sum is unity, leads to

$$p_{Nj} = \frac{e^{-\gamma N - \beta E_j(N)}}{\Xi}, \tag{6.24}$$

where

$$\Xi = \sum_{N=0}^{\infty} \sum_{j} e^{-\gamma N - \beta E_j(N)} = \sum_{N=0}^{\infty} e^{-\gamma N} Z(N) \qquad (6.25)$$

is the grand canonical partition function,[2] and $Z(N)$ is the canonical partition function for $N$ particles.

Substituting the expression for $p_{Nj}$ into the Gibbs entropy

$$S = -k_B \sum_{N=0}^{\infty} \sum_{j} p_{Nj} \ln p_{Nj} , \qquad (6.26)$$

leads to (following similar calculations to those we used in the canonical ensemble)

$$S = k_B \ln \Xi + k_B \beta U + k_B \gamma N. \qquad (6.27)$$

## 6.2.1  Bridge Equation

In Eq. (6.27), $\gamma$ is an undetermined Lagrange multiplier. We can determine it by using similar arguments to those we used for the canonical ensemble. First, consider allowing a small amount of heat to flow from the bath to the system, holding $N$ and $V$ fixed. This leads (as before) to the identification $\beta = 1/k_B T$. Second, suppose that there is a small change $\delta N$ in the number of particles in the system such that the volume $V$ and the internal energy $U$ are held fixed. Under such a change there are small changes in the probabilities of the various microstates, $\delta p_{Nj}$, which satisfy

$$\sum_{N=0}^{\infty} \sum_{j} \delta p_{Nj} = 0; \qquad \sum_{N=0}^{\infty} \sum_{j} \delta p_{Nj} E_j(N) = 0, \qquad (6.28)$$

where the second condition ensures that $U$ does not change. We can now consider the change in the entropy and follow similar steps to those that lead to Eq. (4.15) to get

$$\delta S = k_B \sum_{N=0}^{\infty} \sum_{j} \delta p_{Nj} \left[ \gamma N + \beta E_j(N) \right]$$

$$= k_B \gamma \sum_{N=0}^{\infty} \sum_{j} \delta p_{Nj} N$$

$$= k_B \gamma \delta N, \qquad (6.29)$$

from which we can deduce

$$\frac{\delta S}{\delta N} \longrightarrow \left. \frac{\partial S}{\partial N} \right|_{U,V} = k_B \gamma, \qquad (6.30)$$

---

[2]  To avoid ambiguity we will always refer to the grand canonical partition function using its full name. When we use "partition function" this refers to the partition function in the canonical ensemble. However, not all authors follow this convention and sometimes the grand canonical partition function is referred to as the partition function.

which when we compare to the expression for the chemical potential

$$\mu = -T \left. \frac{\partial S}{\partial N} \right|_{U,V} \tag{6.31}$$

allows us to identify

$$\gamma = -\beta\mu. \tag{6.32}$$

Now that we have related $\gamma$ to the chemical potential, we may write

$$S = k_B \ln \Xi + \frac{U}{T} - \frac{\mu N}{T}, \tag{6.33}$$

which can be rearranged as

$$\Phi = k_B T \ln \Xi = \mu N + TS - U = \mu N - F, \tag{6.34}$$

where $\Phi$ is the grand potential and

$$\Phi = k_B T \ln \Xi \tag{6.35}$$

is the bridge equation for the grand canonical ensemble; the grand canonical partition function is

$$\Xi = \sum_{N=0}^{\infty} e^{\beta \mu N} Z(N), \tag{6.36}$$

with

$$Z(N) = \sum_{j} e^{-\beta E_j(N)}. \tag{6.37}$$

Note that we can view the grand canonical partition function as a weighted average of canonical partition functions, each with different $N$, weighted by the $N^{\text{th}}$ power of the fugacity $z = e^{\beta\mu}$. The probabilities of states in the grand canonical ensemble are proportional to the factors

$$e^{-\beta \mu N - \beta E_j(N)},$$

which are known as Gibbs factors, or generalized Boltzmann factors. If we have states 1 and 2 with $N_1$ and $N_2$ particles and energies $E_1(N_1)$ and $E_2(N_2)$, respectively, then the ratio of their probabilities is given by the ratio of their Gibbs factors:

$$\frac{P(N_1, E(N_1))}{P(N_2, E(N_2))} = \frac{e^{\beta[N_1\mu - E_1(N_1)]}}{e^{\beta[N_2\mu - E_2(N_2)]}}. \tag{6.38}$$

There is less consistency in definitions of the grand potential than the Helmholtz free energy – it is common to see it defined with the opposite sign to our convention, and it is sometimes represented by the symbol $\Omega$. The grand canonical partition function is also sometimes denoted by $\mathcal{Z}$. With our choice of sign convention, $\Phi$ is maximum at equilibrium for fixed $\mu$, $T$ and $V$.

## 6.2.2  Derivatives of the Grand Potential

The three quantities that are fixed in the grand canonical ensemble are $\mu$, $T$ and $V$. Hence if we start from the differential relation for $\Phi$ as

$$d\Phi = N d\mu + \mu dN + SdT + TdS - dU, \tag{6.39}$$

and use the expression Eq. (6.13) that

$$dU = TdS - PdV + \mu dN, \tag{6.40}$$

we obtain

$$d\Phi = SdT + PdV + N d\mu$$
$$= \left.\frac{\partial\Phi}{\partial T}\right|_{\mu,V} dT + \left.\frac{\partial\Phi}{\partial V}\right|_{\mu,T} dV + \left.\frac{\partial\Phi}{\partial\mu}\right|_{T,V} d\mu, \tag{6.41}$$

and we can read off expressions for the quantities $N$, $S$ and $P$:

$$S = \left.\frac{\partial\Phi}{\partial T}\right|_{\mu,V} ; \quad P = \left.\frac{\partial\Phi}{\partial V}\right|_{\mu,T} ; \quad N = \left.\frac{\partial\Phi}{\partial\mu}\right|_{T,V} . \tag{6.42}$$

The internal energy $U$ may be found by rearranging the bridge equation:

$$U = \mu N + TS - \Phi. \tag{6.43}$$

Another thermodynamic potential of interest is the Gibbs free energy

$$G = U + PV - TS, \tag{6.44}$$

which is minimized at equilibrium for constant $P$ and $T$. It is straightforward to show that

$$dG = \mu dN + VdP - SdT, \tag{6.45}$$

and hence

$$\mu = \left.\frac{\partial G}{\partial N}\right|_{P,T} . \tag{6.46}$$

In many cases it is true that $G = \mu N$ (provided $\mu$ is independent of $N$).

## Average Number of Particles

The average number of particles in the system, $N$, can be determined using the grand canonical partition function as follows:

$$N = \langle N \rangle = \frac{\sum_{N=0}^{\infty} \sum_j N e^{\beta\mu N - \beta E_j(N)}}{\Xi}$$

$$= \frac{1}{\beta} \frac{1}{\Xi} \frac{\partial \Xi}{\partial \mu}$$

$$= \frac{1}{\beta} \frac{\partial \ln \Xi}{\partial \mu}$$

$$= \frac{\partial \Phi}{\partial \mu}, \tag{6.47}$$

which is identical to the relation we obtained from the thermodynamic potential in Eq. (6.42).

## Number Fluctuations

We saw in the canonical ensemble that it was possible to relate fluctuations in the mean energy to the specific heat (the rate of change of the energy in temperature). We can see a similar relationship in the grand canonical ensemble if we look at the average particle number:

$$\frac{\partial \mathcal{N}}{\partial \mu} = \frac{\partial \langle N \rangle}{\partial \mu} = \frac{\partial}{\partial \mu} \left[ \frac{\sum_{N=0}^{\infty} \sum_j N e^{\beta \mu N - \beta E_j(N)}}{\Xi} \right]$$

$$= -\frac{1}{\Xi^2} \frac{\partial \Xi}{\partial \mu} \sum_{N=0}^{\infty} \sum_j N e^{\beta \mu N - \beta E_j(N)} + \beta \frac{\sum_{N=0}^{\infty} \sum_j N^2 e^{\beta \mu N - \beta E_j(N)}}{\Xi}$$

$$= \frac{\langle N^2 \rangle - \langle N \rangle^2}{k_B T}$$

$$= \frac{\sigma_N^2}{k_B T}, \tag{6.48}$$

where we used $\langle N \rangle = \frac{1}{\beta} \frac{1}{\Xi} \frac{\partial \Xi}{\partial \mu}$. We see that the rate of change of $\mathcal{N}$ with $\mu$ is proportional to the number fluctuations at equilibrium. We can view the quantity $\partial \mathcal{N}/\partial \mu$ as a "number compressibility", which indicates how responsive the mean number of particles is to changes in chemical potential. There are examples of quantum systems which are "incompressible", i.e. $\partial \mathcal{N}/\partial \mu = 0$ at low temperatures, such as quantum Hall states or Mott insulators. These states emerge when electrons have strong interactions with each other and are beyond the scope of our discussions here.

# 6.3  Examples

We now consider several examples which illustrate the application of the grand canonical ensemble. We will also utilize the grand canonical ensemble further in Chapter 7.

## 6.3.1  Fermions in a Two-Level System

We previously considered the example of a two-level system in the canonical ensemble in Section 4.4.1, at the one-particle level, when we calculated $Z_1$. Now we consider a two-level system with energy levels 1 and 2, which have energies $\epsilon_1 = 0$ and $\epsilon_2 = \epsilon$, respectively. If we consider fermions in this two-level system, then the Pauli exclusion principle[3] ensures that there can be at most one particle per energy level. Hence, when we calculate the grand canonical partition function, we only need to consider $N = 0, 1$ or $2$. The respective allowed energies for each $N$ are

$$E(0) = 0,$$
$$E(1) = 0 \text{ or } \epsilon,$$
$$E(2) = \epsilon.$$

Writing the grand canonical partition function as

$$\Xi = \sum_{N=0}^{2} e^{\beta \mu N} Z(N), \qquad (6.49)$$

with

$$Z(N) = \sum_{s} e^{-\beta E_s(N)}, \qquad (6.50)$$

we can calculate straightforwardly that

$$Z(0) = 1,$$
$$Z(1) = 1 + e^{-\beta \epsilon},$$
$$Z(2) = e^{-\beta \epsilon},$$

from which we obtain

$$\Xi = 1 + e^{\beta \mu} \left(1 + e^{-\beta \epsilon}\right) + e^{2\beta \mu - \beta \epsilon}$$
$$= \left(1 + e^{\beta \mu}\right) \left(1 + e^{\beta \mu - \beta \epsilon}\right). \qquad (6.51)$$

We can calculate the average energy by using

$$U = \frac{1}{\Xi} \sum_{N=0}^{2} \sum_{j} E_j(N) e^{\beta \mu N - \beta E_j(N)}, \qquad (6.52)$$

which gives

$$U = \frac{\epsilon\, e^{\beta \mu - \beta \epsilon} + \epsilon\, e^{2\beta \mu - \beta \epsilon}}{\left(1 + e^{\beta \mu}\right)\left(1 + e^{\beta \mu - \beta \epsilon}\right)}$$
$$= \frac{\epsilon\, e^{\beta \mu - \beta \epsilon}}{1 + e^{\beta \mu - \beta \epsilon}}$$
$$= \frac{\epsilon}{e^{\beta(\epsilon - \mu)} + 1}. \qquad (6.53)$$

We can also calculate the average number of fermions in the two-level system using Eq. (6.47), and after a few lines of algebra this gives

$$N = \frac{1}{e^{-\beta \mu} + 1} + \frac{1}{e^{\beta(\epsilon - \mu)} + 1}. \qquad (6.54)$$

[3] We discuss the Pauli exclusion principle in more detail in Chapter 7.

Note that in both the expressions for $U$ and for $\mathcal{N}$, Eqs (6.53) and (6.54), respectively, there are terms in the denominator involving exponentials of $\beta\mu$ and $\beta\epsilon$. These are examples of the Fermi–Dirac distribution which we will study in greater detail in Chapters 7 and 8.

## 6.3.2 The Langmuir Adsorption Isotherm

In an adsorption process a vapour of some species is in equilibrium with molecules of the vapour that are bound to a surface. Suppose that there are $N_s$ sites on the surface at which molecules from the vapour can bind, and that when they bind they lower their energy by an amount $\epsilon$. We can ignore vibrations for the molecules in the vapour, but for molecules bound to the surface, the energy scale of vibrations is lower, and so we must include them in the calculation of the partition function. We label the partition function associated with vibrations of molecules on the surface as $z_s$ and assume that there are $N$ surface sites occupied by molecules. We will calculate the relationship between the vapour pressure and the fraction of sites on the surface that are occupied: $f = N/N_s$.

Each of the surface sites are independent, so we can consider the grand canonical partition function for a single surface site, since the probability of a site being occupied is the same for every site, noting that the energy of a bound state is $-\epsilon$:

$$\Xi = \sum_{N=0}^{1} \sum_{j} e^{\beta\mu N - \beta E_j(N)} = 1 + z_s e^{\beta(\mu+\epsilon)}. \tag{6.55}$$

There are two choices, either the molecule is bound or not bound, so there are a total of two terms in the partition function.

We assume that the system is in equilibrium, hence the chemical potential for the molecules on the surface is the same as for the molecules in the vapour phase, i.e. $\mu_{\text{surface}} = \mu_{\text{vapour}} = \mu$. From the grand canonical partition function, Eq. (6.55), we can obtain the grand potential

$$\Phi = k_B T \ln \Xi = k_B T \ln\left(1 + z_s e^{\beta(\mu+\epsilon)}\right), \tag{6.56}$$

from which we can obtain the fraction of sites that are occupied, which is given by

$$\begin{aligned}
f &= \left.\frac{\partial \Phi}{\partial \mu}\right|_{T,V} \\
&= k_B T \frac{\partial}{\partial \mu} \ln\left(1 + z_s e^{\beta(\mu+\epsilon)}\right) \\
&= k_B T \beta \frac{z_s e^{\beta(\mu+\epsilon)}}{1 + z_s e^{\beta(\mu+\epsilon)}} \\
&= \frac{1}{\frac{1}{z_s} e^{-\beta(\epsilon+\mu)} + 1}.
\end{aligned} \tag{6.57}$$

Treating the vapour as an ideal gas, we can relate $\mu$ to the pressure by using

$$P = k_B T n_Q Z_r e^{\beta\mu}, \tag{6.58}$$

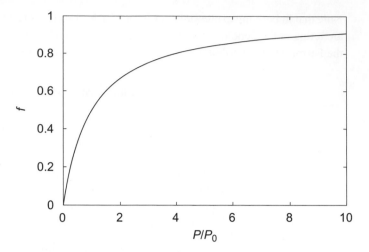

Fig. 6.3 Langmuir adsorption isotherm: fractional adsorption $f$ as a function of pressure $P/P_0$.

which allows us to write

$$f = \frac{1}{\frac{k_B T n_Q Z_r}{P z_s} e^{-\beta \epsilon} + 1} = \frac{P}{P_0 + P}, \tag{6.59}$$

where we define

$$P_0 = \frac{k_B T n_Q Z_r}{z_s} e^{-\beta \epsilon} \tag{6.60}$$

as a characteristic pressure scale for the system.

This is an example of a *Langmuir adsorption isotherm* (Fig. 6.3), where the scale of the curve is determined by $P_0$ once $T$ is fixed. As the pressure $P$ of the gas increases relative to $P_0$, the fraction of molecules adsorbed tends to 1. In addition to the generic example of molecules being adsorbed on a surface considered here, numerous biological processes can be understood as examples of adsorption. In particular, the uptake of oxygen by myoglobin and hemoglobin molecules in muscles and blood respectively, can be described in a similar way to the preceding discussion.

## 6.4 Chemical Equilibrium and the Law of Mass Action

Suppose we have a system containing $m$ different types of molecule with chemical symbols $B_1, B_2, \ldots, B_m$, which can undergo chemical reactions in which some set of molecules transform into a different set of molecules. Such chemical reactions must be consistent with the total number of atoms of each element present. A simple example is the chemical equilibrium between hydrogen, oxygen and water molecules in the gas phase:

$$2H_2 + O_2 \rightleftharpoons 2H_2O. \tag{6.61}$$

Let $b_i$, which is some integer, be the coefficient of $B_i$ in a chemical equation, then we consider $b_i$ positive for products of a reaction and $b_i$ negative for reactant molecules that get depleted in a reaction. We can then write a general chemical equation in the form

$$\sum_{i=1}^{m} b_i B_i = 0, \tag{6.62}$$

which for the reaction Eq. (6.61) would correspond to

$$-2H_2 - O_2 + 2H_2O = 0, \tag{6.63}$$

with $b_1 = -2$, $b_2 = -1$, $b_3 = 2$, $B_1 = H_2$, $B_2 = O_2$ and $B_3 = H_2O$.

If we let $N_i$ be the number of $B_i$ molecules in the system, then the change in each species as the reaction progresses must be proportional to $b_i$, so for all $i$, $dN_i = \lambda b_i$ for some constant $\lambda$ and $dN_i > 0$ for reaction products and $dN_i < 0$ for reactants that disappear in the reaction.

The discussion above has not made any reference to equilibrium: if we consider a situation in which the molecules are in a container with fixed volume and temperature,[4] then the condition that the Helmholtz free energy is minimized implies $dF = 0$, which is equivalent to[5]

$$dF = \sum_{i=1}^{m} \left. \frac{\partial F}{\partial N_i} \right|_{T,V} dN_i = \sum_{i=1}^{m} \mu_i dN_i = 0, \tag{6.64}$$

where

$$\mu_i = \left. \frac{\partial F}{\partial N_i} \right|_{T,V,N_j \neq N_i} \tag{6.65}$$

is the chemical potential for species $i$ (note that the $N_j$ for all species with $j \neq i$ are held constant in determining $\mu_i$). The condition that all of the $dN_i$ satisfy the chemical equation implies that

$$\sum_{i=1}^{m} b_i \mu_i = 0, \tag{6.66}$$

which is the general condition for chemical equilibrium.

## 6.4.1 The Law of Mass Action

Consider a chemical reaction of the form

$$\nu_A A + \nu_B B \rightleftharpoons \nu_C C + \nu_D D. \tag{6.67}$$

---

[4] Chemists or those working with biological systems tend to prefer to work with constant pressure rather than constant volume, which corresponds to the Gibbs ensemble, in which case the Gibbs free energy $G$ rather than the Helmholtz free energy $F$ is minimized at equilibrium.

[5] Note that the $\mu_i$s don't have to be equal in equilibrium because we are not considering the system in the grand canonical ensemble, but instead are using the Helmholtz free energy, in which case $F$ is minimized, not $\Phi$ (which *would* imply $\mu$ constant).

It is possible to establish a relationship between the concentrations of reactants, which define an *equilibrium constant*. The equilibrium constant is

$$K = \frac{n_C^{\nu_C} n_D^{\nu_D}}{n_A^{\nu_A} n_B^{\nu_B}},$$

(6.68)

where $n_i$ is the concentration of atoms of species $i$. We can generalize to $m$ species and then

$$K = \prod_{i=1}^{m} n_i^{b_i},$$

(6.69)

in the notation we used above. We now derive Eq. (6.69) from our knowledge of chemical equilibrium.

If we have a set of molecules in equilibrium we might want to know the relationship between the mean numbers of different molecules. We can address this by considering the Helmholtz free energy of the system at constant volume and temperature. If $b_i$ of each of the reactant molecules are transformed into $b_j$ of each of the product molecules, then

$$\Delta F = \lambda \sum_i \frac{\partial F}{\partial N_i}\bigg|_{T,V} b_i = \lambda \sum_i \mu_i b_i.$$

(6.70)

In equilibrium we know from Eq. (6.66) that $\sum_i \mu_i b_i = 0$, so $\Delta F = 0$.

If we focus on the situation in which all of the molecules are in the gas phase and ignore the effects of each species on the others, i.e. we take the dilute reactants limit, then the chemical potential for each species is (using the result we found for an ideal gas)

$$\mu_i = k_B T \ln\left(\frac{N_i}{g_s^i n_q^i V Z_{\text{int}}^i}\right),$$

(6.71)

where $g_s^i$ is the spin degeneracy of species $i$, $n_Q^i$ the quantum concentration for species $i$ and $Z_{\text{int}}^i$ the partition function for internal degrees of freedom (i.e. rotations, vibrations, electronic degrees of freedom) in the canonical ensemble for species $i$.

Hence, when we substitute Eq. (6.71) for each species into Eq. (6.70) we get

$$\sum_{i=1}^{m} k_B T b_i \ln(N_i) = \sum_{i=1}^{m} b_i k_B T \ln(g_s^i n_Q^i V Z_{\text{int}}^i),$$

(6.72)

and, exponentiating both sides of the equation,

$$\prod_{i=1}^{m} N_i^{b_i} = \prod_{i=1}^{m} \left(g_s^i n_Q^i V\right)^{b_i} e^{b_i \ln Z_{\text{int}}^i},$$

(6.73)

and dividing both sides of the equation by $V^{\sum_{i=1}^{m} b_i}$, we obtain an expression for the equilibrium constant:

$$K = \prod_{i=1}^{m} n_i^{b_i} = \prod_{i=1}^{m} \left(g_s^i n_Q^i\right)^{b_i} e^{-\beta b_i F_{\text{int}}^i},$$

(6.74)

where

$$F_{\text{int}}^i = -k_B T \ln\left(Z_{\text{int}}^i\right).$$ 

(6.75)

Equation (6.74) shows the form of the equilibrium constant $K$ for a chemical reaction and is known as the **law of mass action**. For a particular chemical reaction, $K$ is a constant at fixed temperature $T$. The significance of this result is that if we know the internal partition functions for the species involved in a chemical reaction, we can determine the relation between the number of each species when the system is at equilibrium.

Chemical reactions are often not carried out at constant volume, but at constant pressure. Similar arguments to those above can be made for constant pressure, but one should use the Gibbs free energy in place of the Helmholtz free energy.

## 6.5 Summary

In the grand canonical ensemble, the temperature $T$, chemical potential $\mu$ and volume $V$ are fixed and energy and particle number may fluctuate. The equation

$$\Phi = k_B T \ln \Xi$$

(6.76)

is the bridge equation, which relates the grand potential

$$\Phi = \mu N + TS - U = \mu N - F$$

(6.77)

to the grand canonical partition function

$$\Xi = \sum_N e^{\beta \mu N} Z(N),$$

(6.78)

which is a weighted sum over partition functions for the canonical ensemble with $N$ particles:

$$Z(N) = \sum_s e^{-\beta E_s(N)}.$$

(6.79)

# Problems

**6.1**  Show that for a monatomic ideal gas the fugacity $z = n/n_Q$.

**6.2**  We can calculate the variation of pressure in an isothermal atmosphere by equating the chemical potential at heights 0 and $h$. Use this argument to show that the pressure at height $h$, $p(h)$, is

$$p(h) = p(0)e^{-\frac{mgh}{k_B T}}.$$

This result is not applicable to the Earth's atmosphere as it is not isothermal.

**6.3**  Consider an ultra-relativistic ideal gas (where we can ignore the rest mass of the particles), for which the energies of the states are given by

$$E = |\mathbf{p}|c.$$

Calculate the single-particle classical partition function in the canonical ensemble

$$Z_1 = \frac{1}{(2\pi\hbar)^3} \int d^3\mathbf{x} \int d^3\mathbf{p}\, e^{-\beta E},$$

and use this to obtain an expression for the chemical potential of an $N$-particle system.

**6.4**  Consider a two-level system with energy levels 0 and $\epsilon$ populated by bosons (bosons have no restrictions on how many particles can be in each energy level). Find the average energy and occupation of the levels and calculate the heat capacity.

**6.5**  Hydrogen atoms may be adsorbed from a vapour phase onto the surface of liquid helium, where they behave as a weakly interacting two-dimensional gas. Take the mass of the hydrogen atoms to be $m$, their vapour pressure to be $P$ and their adsorption energy to be $\Delta$. Treat the hydrogen vapour as a particle reservoir for the gas on the surface and treat both the two- and three-dimensional gases as ideal gases. Hence show that the number density per unit area of hydrogen atoms on the surface is

$$n_{2d} = \left(\frac{P}{k_B T}\right)\left(\frac{2\pi\hbar^2}{mk_B T}\right)^{\frac{1}{2}} e^{\beta\Delta}.$$

**6.6**  Consider a gas of non-interacting atoms with magnetic moment $\mathbf{m}$ that are placed in a magnetic field $\mathbf{B}$. Assume that the atoms are either spin up ($\uparrow$), in which case their magnetic energy $E = -mB$, or spin down ($\downarrow$), for which the energy is $E = mB$.

  (a) Find the classical one-particle canonical partition function for spin-up particles and then do the same for spin-down particles.
  (b) Find the chemical potentials $\mu_\uparrow$ and $\mu_\downarrow$ for the $N_\uparrow$ up spins and $N_\downarrow$ down spins. Assume that any up spin is indistinguishable from any other up spin, and similarly for down spins.
  (c) Suppose $\mathbf{B}$ is a slowly varying function of $x$. Consider

$$\frac{\partial \mu_\uparrow}{\partial x} \quad \text{and} \quad \frac{\partial \mu_\downarrow}{\partial x},$$

noting that the up and down spins are in equilibrium with each other, to determine

$$n_\uparrow = \frac{1}{2} n_0 e^{\frac{mB}{k_B T}} \quad \text{and} \quad n_\downarrow = \frac{1}{2} n_0 e^{-\frac{mB}{k_B T}},$$

where $n_0$ is the density at a location where $B = 0$.

(d) Use part (c) to find the total density as a function of magnetic field, and show that for small fields ($k_B T \gg mB$)

$$n(B) \simeq n_0 \left( 1 + \frac{m^2 B^2}{2 k_B^2 T^2} \right).$$

**6.7**  Invert Eq. (6.54) to find the chemical potential as a function of $\mathcal{N}$, $k_B T$ and $\epsilon$ in a fermionic two-level system.

**6.8**  Consider a mixture of $N_A$ atoms of type $A$ and $N_B$ atoms of type $B$ in a volume $V$, and use the notation

$$X_A = \frac{N_A}{N_A + N_B}; \qquad X_B = \frac{N_B}{N_A + N_B}.$$

(a) Treating component $A$ as an ideal gas, show that

$$\mu_A = -k_B T \ln \left[ \left( \frac{m_A k_B T}{2\pi\hbar^2} \right)^{\frac{3}{2}} \frac{g_A V}{N_A + N_B} \right] + k_B T \ln X_A = \mu_A^0 + k_B T \ln X_A,$$

where $g_A$ is the spin degeneracy for $A$ atoms.

The term $\mu_A^0$ is the chemical potential we would have if all atoms were of type $A$, and only the second term in the expression for $\mu_A$ refers to the concentration of $A$ particles.

(b) By neglecting interactions between particles, evaluate the multiplicity function for the number of ways to arrange particles $A$ and $B$ in the gas, and determine the entropy of mixing $S_{\text{mixing}}$ (i.e. the excess entropy beyond that of $N_A$ atoms of $A$ and $N_B$ atoms of $B$ that are unmixed) that this implies. Hence show that the Gibbs free energy may be written as

$$G = N_A \mu_A^0 + N_B \mu_B^0 - T S_{\text{mixing}}.$$

In a liquid it is unrealistic to expect that interactions are unimportant, however, if we assume that mixing is not greatly affected by interactions, then we may write for component $\alpha$:

$$\mu_\alpha(P, T, X_\alpha) = \mu_\alpha^0(P, T) + k_B T \ln X_\alpha,$$

in which we take $\mu_\alpha^0(P, T)$ to be the chemical potential of pure component $\alpha$ at the same temperature $T$ and pressure $P$ as in our mixture. Our ignorance of interaction effects has been placed in $\mu_\alpha^0(P, T)$ and we cannot assume it has the form we would infer for an ideal gas as in part (a).

(c) We allow our mixture to have both a liquid and a vapour component. If the liquid and vapour components of $A$ are in equilibrium, what does this imply for $\mu_A(\text{liquid})$ and $\mu_A(\text{vapour})$? The same result will hold for component $B$.

We can write explicit expressions for $\mu_A(\text{vapour})$ and $\mu_B(\text{vapour})$ if we assume that they behave as non-interacting ideal gases which satisfy Dalton's law of partial pressures

$$PV = (N_A + N_B)k_B T = (P_A + P_B)V,$$

where $P_A$ and $P_B$ are the partial pressures for components $A$ and $B$, respectively. In the limit that either $A$ or $B$ are pure liquids, these partial pressures take the values $P_A^0$ and $P_B^0$. Hence, $P_A$ and $P_B$ are functions of concentration and temperature, but $P_A^0$ and $P_B^0$ depend only on temperature.

(d) Determine an expression for $\mu_A(\text{vapour})(P_A, T)$ making use of the considerations above.

(e) In general we do not know what form to assume for $\mu_\alpha^0(\text{liquid}, P, T)$. Assume that $P_A^0$ (the vapour pressure of pure liquid $A$) is known and use your answer to part (c) to determine $\mu_A^0(\text{liquid}, P_A^0, T)$ for the liquid phase.

(f) The difference between $\mu_\alpha^0(\text{liquid}, P, T)$ and $\mu_\alpha^0(\text{liquid}, P_A^0, T)$ is typically small for moderate pressures, so we can obtain an approximate expression for $\mu_A^0(\text{liquid}, P, T)$ by making a Taylor expansion:

$$\mu_A^0(\text{liquid}, P, T) \simeq \mu_A^0(\text{liquid}, P_A^0, T) + (P - P_A^0)\left(\frac{\partial \mu_A^0}{\partial P}\right)_T.$$

Use the Gibbs–Duhem relation $Nd\mu = VdP - SdT$ to determine that

$$\mu_A^0(\text{liquid}, P, T) \simeq \mu_A^0(\text{liquid}, P_A^0, T) + (P - P_A^0)v_A^0,$$

where $v_A^0$ is the volume per atom (which we assume not to vary with $P$), and hence determine that

$$\mu_A(\text{liquid}, P, X_A) = -k_B T \ln\left[\left(\frac{m_A k_B T}{2\pi\hbar^2}\right)^{\frac{3}{2}} \frac{g_A k_B T}{P_A^0}\right] + (P - P_A^0)v_A^0 + k_B T \ln X_A,$$

and similarly for component $B$.

(g) Using the previously obtained results (especially part (c)), show that

$$P = P_A + P_B = X_A P_A^0 \exp\left\{\frac{(P - P_A^0)v_A^0}{k_B T}\right\} + X_B P_B^0 \exp\left\{\frac{(P - P_B^0)v_B^0}{k_B T}\right\}.$$

In the limit that the exponential factors can be approximated by unity, this result simplifies to Raoult's law:

$$P = X_A P_A^0 + X_B P_B^0.$$

(h) At $T = 115$ K the vapour pressure of liquid argon is $9.4\times10^5$ Pa and its density is $1.07\times10^3$ kg m$^{-3}$. At the same temperature the vapour pressure of liquid krypton is $6.9\times10^4$ Pa and its density is $2.34\times10^3$ kg m$^{-3}$. Do you find Raoult's law to be valid for a liquid mixture of 30% argon atoms and 70% krypton atoms at 115 K?

# 7 Quantum Statistical Mechanics

In Chapter 4, when discussing distinguishable and indistinguishable particles, we saw that the available microstates for different types of quantum particles (fermions and bosons) are not identical. This should be expected to have statistical mechanical consequences. To this point we have only been interested in the classical, dilute limit in which there is at most one particle per energy level. In order to investigate how the quantum nature of particles affects their statistical mechanical properties, we first review how the quantum statistics of fermions and bosons arise. We then use the grand canonical ensemble to calculate the grand canonical partition functions for fermions and bosons, and hence the Fermi–Dirac and Bose–Einstein distributions. In Chapters 8 and 9 we discuss the properties of fermions and bosons, respectively, in more detail.

## 7.1 Quantum Statistics

As we discussed in Chapter 4, quantum particles are indistinguishable – there is no way to distinguish one electron from any other electron, for instance. This is in sharp contrast to classical particles which are distinguishable. If we know the initial positions and momenta of classical particles, then in principle we can determine their positions and momenta at all future times.

To illustrate these ideas, consider the situation pictured in Fig. 7.1, in which we have either three identical classical particles or three identical quantum particles $A$, $B$ and $C$, each initially prepared in some well-characterized initial state. If we know the Hamiltonian we can use the initial state information of position and momenta to determine the trajectories of the three classical particles and distinguish them at any later time. For the quantum particles, if we make a measurement at some later time $t$ and find particles at the locations initially occupied by $A$, $B$ and $C$, we have no way of knowing whether the particle we measure is indeed the particle that was there initially or not, so we should treat the particles as indistinguishable.

What does this mean at the level of the wavefunction $\psi$? If we let the initial co-ordinates of particles $A$, $B$ and $C$ be $x_A$, $x_B$ and $x_C$, respectively, then at time $t$ the wavefunction is a function of the form

$$\psi(x_A, x_B, x_C).$$

However, as there is no way to tell whether it is particle $A$ at $x_A$ or $x_B$, we should take into account the $3! = 6$ ways to arrange the particles. This means we should also

Classical particles

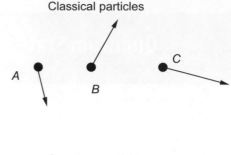

Quantum particles

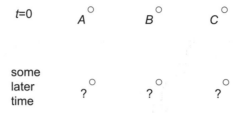

Fig. 7.1 Distinguishable classical particles and indistinguishable quantum particles. The trajectories of the classical particles are determined once their initial conditions are known. We only have information about the quantum particles when we make measurements.

consider $\psi(x_B, x_A, x_C)$, and all other pairwise exchanges of particles. Since the particles are identical, the physical observables associated with the wavefunction in which particle $A$ is at $x_A$ and particle $B$ is at $x_B$ and the wavefunction in which particle $B$ is at $x_A$ and particle $A$ is at $x_B$ are identical. Hence the probability density $\rho = \psi^*\psi = |\psi|^2$ is unaffected.

If $\psi_1$ corresponds to one ordering of the particles, e.g. $\psi(x_A, x_B, x_C)$, and $\psi_2$ corresponds to another ordering of the particles, e.g. $\psi(x_B, x_C, x_A)$, then given that we cannot distinguish these two possibilities, we must have

$$\rho = |\psi_1|^2 = |\psi_2|^2, \tag{7.1}$$

which implies

$$\psi_1 = e^{i\theta}\psi_2, \tag{7.2}$$

for some $\theta$, i.e. the wavefunctions of two orderings of particles which are physically indistinguishable are the same up to a phase. Equivalently we can view this as the wavefunction acquiring a phase if we interchange two particles. In three dimensions it has been proven (this is the *spin-statistics theorem*) that the only allowed values for $\theta$ are 0 and $\pi$, with $\theta = 0$ for bosons and $\theta = \pi$ for fermions, respectively.[1] For pairwise exchange of particles,

---

[1]  In two dimensions it is possible to have values of $\theta$ other than 0 or $\pi$ which correspond to particles with so-called "fractional statistics" or anyons. Anyons emerge in the theoretical description of the fractional quantum Hall effect as excitations that arise from the interactions of many electrons.

$$\psi(x_B, x_A, x_C) = \begin{cases} \psi(x_A, x_B, x_C), & \text{bosons} \\ -\psi(x_A, x_B, x_C), & \text{fermions} \end{cases}. \qquad (7.3)$$

Bosons have integer spin, and examples include: elementary particles such as photons, W and Z bosons and the Higgs boson; collective excitations of solids such as magnons and phonons; and composite objects such as mesons or atoms, including $^4$He. Fermions have half odd-integer spin: electrons, muons, tau leptons, neutrinos and quarks are all fermionic elementary particles, as are atoms such as $^3$He.

The antisymmetry of fermion wavefunctions under particle exchange has important consequences. To see this, suppose $x_A = x_B = x$, then we have.

$$\psi(x, x, x_C) = -\psi(x, x, x_C) = 0, \qquad (7.4)$$

so we cannot have two fermions occupying the same state. This is known as the **Pauli exclusion principle**. We have illustrated the exclusion principle for two or three fermions, but it also holds for any number of fermions. If we have $N$ fermions, then the wavefunction is antisymmetric under interchange of any two of the $N$ fermions.

## 7.2 Distinguishable Particles and Maxwell–Boltzmann Statistics

Unless otherwise specified, in Chapters 7, 8 and 9 we will deal with non-interacting particles. In a non-interacting picture we start with orbitals, e.g. $s$, $p$, $d$ ... states in a hydrogen atom, and populate them with particles. The energy of an $N$-particle state is obtained by simply adding the energies of each of the orbitals, $\epsilon_s$, multiplied by their occupations, $n_s$, so that

$$E[\{n_s\}] = \sum_s n_s \epsilon_s. \qquad (7.5)$$

In general, for interacting systems there will be additional contributions to the energy that come from multiple particles occupying the same orbital. However, for sufficiently weakly interacting systems we can accurately approximate the $N$-particle state by a non-interacting quantum state.

We showed in Eq. (4.75) that the partition function of $N$ distinguishable non-interacting particles can be written simply as $Z = (Z_1)^N$, where $Z_1$ is the single-particle partition function. For $N$ indistinguishable particles there is no general simple closed form for the partition function. In order to overcome this barrier to calculation in the case of an ideal gas, we investigated the dilute limit of indistinguishable particles (for which there is no double occupancy of states, so the distinction between fermions and bosons is irrelevant). This allowed us to relate the partition function for $N$ dilute indistinguishable particles to the partition function for $N$ distinguishable particles, which can be expressed simply in terms of $Z_1$, the single-particle partition function, as

$$Z \simeq \frac{(Z_1)^N}{N!}. \qquad (7.6)$$

Our subsequent calculations from this partition function reproduced the ideal gas law, which we understood to be the classical limit of a quantum gas. However, we know that classical particles are actually distinguishable, which would naively imply that $Z = (Z_1)^N$ with no factor of $N!$. This leads to different predictions for the physical properties of an ideal gas that do not agree with experiment. How do we resolve this apparent contradiction?

Whilst in principle we could follow the trajectories of every particle in an ideal gas, in practice we do not and our theoretical formalism (i.e. the Hamiltonian) treats all atoms in the classical gas identically. This leads to the idea of *undistinguished* particles – if we take two undistinguished particles, $A$ and $B$, then the phase-space points $(\mathbf{p}_A, \mathbf{p}_B, \mathbf{q}_A, \mathbf{q}_B)$ and $(\mathbf{p}_B, \mathbf{p}_A, \mathbf{q}_B, \mathbf{q}_A)$ should not both be counted, since the Hamiltonian is symmetric under the interchange of $A$ and $B$. This halves the volume of phase space compared to two distinguishable particles. If we generalize to $N$ particles, then we should reduce the volume of phase space by a factor of $N!$, which is the factor that enters into the partition function. This factor was originally introduced by Gibbs to obtain physically sensible behaviour for the entropy of mixing. In the classical context this has a slightly *ad hoc* feeling about it, whereas if one takes the view that even classical particles are really quantum, then the connection to the dilute limit of indistinguishable particles is a more satisfying explanation.

### 7.2.1  Maxwell–Boltzmann Statistics

Despite any misgivings we may have regarding undistinguished particles, we can go ahead and see what is implied for their statistical mechanics. We will use the grand canonical ensemble to perform calculations. Rather than treating the particles as independent systems, we will treat the energy levels as independent systems. The particles do not interact, hence the total energy associated with each energy level is simply the product of its energy and its occupation, which may fluctuate in equilibrium. With this viewpoint, it is natural to work in the grand canonical ensemble. The grand canonical partition function is

$$\Xi = \sum_{N=0}^{\infty} e^{\beta \mu N} Z(N), \tag{7.7}$$

where $Z(N)$ is the canonical partition function for $N$ particles. For undistinguished particles, we start from the form of the partition function in Eq. (7.6) to obtain

$$\Xi = \sum_{N=0}^{\infty} e^{\beta \mu N} \frac{Z_1^N}{N!}$$

$$= \sum_{N=0}^{\infty} \frac{\left(e^{\beta \mu} Z_1\right)^N}{N!}$$

$$= e^{e^{\beta \mu} Z_1}. \tag{7.8}$$

The single-particle partition function is

$$Z_1 = \sum_s e^{-\beta \epsilon_s}, \tag{7.9}$$

and hence the grand canonical partition function is

$$
\begin{aligned}
\Xi &= e^{e^{\beta \mu} \sum_s e^{-\beta \epsilon_s}} \\
&= \prod_s e^{e^{\beta(\mu-\epsilon_s)}},
\end{aligned}
\tag{7.10}
$$

so the grand potential is

$$
\Phi = k_B T \ln \Xi = k_B T \sum_s e^{\beta(\mu-\epsilon_s)},
\tag{7.11}
$$

from which we can determine

$$
N = \sum_s \langle n_s \rangle = \frac{\partial \Phi}{\partial \mu} = \sum_s e^{\beta(\mu-\epsilon_s)},
\tag{7.12}
$$

and hence

$$
\langle n_s \rangle = e^{\beta(\mu-\epsilon_s)},
\tag{7.13}
$$

which is known as the Maxwell–Boltzmann distribution, and the fictitious particles (that are neither fermion nor boson) that have the properties described above are said to follow Maxwell–Boltzmann statistics.

## 7.3  Quantum Particles in the Grand Canonical Ensemble

In Chapter 4 we discussed the partition function for fermions and bosons in the canonical ensemble, in which case the partition function has to be evaluated subject to the constraint that $N = \sum_s n_s$. It is much less mathematically taxing to treat quantum particles in the grand canonical ensemble. In the grand canonical ensemble we may write the grand canonical partition function as (where $\{n_s\}$ is a set of occupation numbers for the states labelled by $s$)

$$
\begin{aligned}
\Xi &= \sum_{N=0}^{\infty} e^{\beta \mu N} \sideset{}{'}\sum_{\{n_s\}}^{X} \prod_s e^{-\beta n_s \epsilon_s} \\
&= \sum_{N=0}^{\infty} \sideset{}{'}\sum_{\{n_s\}}^{X} e^{\beta \mu N} \prod_s e^{-\beta n_s \epsilon_s} \\
&= \sum_{\{n_s\}}^{X} \prod_s e^{\beta(\mu-\epsilon_s)n_s} \\
&= \prod_s \sum_{n_s=0}^{X} e^{\beta(\mu-\epsilon_s)n_s}.
\end{aligned}
\tag{7.14}
$$

Note that the sum over $\{n_s\}$ in the first two lines of Eq. (7.14) has the constraint that $\sum_s n_s = N$, indicated by a prime, and the upper limit on the values of $n_s$, $X$, is determined by the particle statistics ($X = 1$ for fermions and $X = \infty$ for bosons). The rewriting of the sum is a general result for non-interacting particles that the grand canonical partition function can be written as a product of the individual grand canonical partition functions for each distinct energy level. We used this property of the grand canonical partition function when determining the Langmuir adsorption isotherm in Section 6.3.2.

## Fermions

For fermions, $X = 1$ and there are only two terms in the sum over $n_s$, which immediately leads to the grand canonical partition function:

$$\Xi_f = \prod_s \left[1 + e^{\beta(\mu-\epsilon_s)}\right].\tag{7.15}$$

## Bosons

For bosons, $X = \infty$, so the sum over $n_s$ becomes an infinite geometric series:

$$\sum_{n_s=0}^{X} e^{\beta(\mu-\epsilon_s)n_s} = \sum_{n_s=0}^{\infty} e^{\beta(\mu-\epsilon_s)n_s} = \frac{1}{1 - e^{\beta(\mu-\epsilon_s)}},\tag{7.16}$$

and hence the grand canonical partition function is

$$\Xi_b = \prod_s \frac{1}{1 - e^{\beta(\mu-\epsilon_s)}}.\tag{7.17}$$

We can combine the results in Eqs (7.15) and (7.17) to allow us to derive results for both fermions and bosons in parallel:

$$\Xi_{fb} = \prod_s \left[1 \pm e^{\beta(\mu-\epsilon_s)}\right]^{\pm 1},\tag{7.18}$$

where the $+$ sign refers to fermions and the $-$ sign refers to bosons.

We can evaluate the grand potential for both fermions ($f$) and bosons ($b$):

$$\Phi_{fb} = k_B T \ln \Xi_{fb} = \pm k_B T \sum_s \ln\left[1 \pm e^{\beta(\mu-\epsilon_s)}\right].\tag{7.19}$$

The average particle number is determined by

$$N = \left.\frac{\partial \Phi}{\partial \mu}\right|_{V,T},\tag{7.20}$$

hence

$$N = \pm k_B T \sum_s \frac{1}{1 \pm e^{\beta(\mu-\epsilon_s)}}\left[\pm\frac{e^{\beta(\mu-\epsilon_s)}}{k_B T}\right]$$

$$= \sum_s \frac{1}{e^{\beta(\epsilon_s-\mu)} \pm 1},\tag{7.21}$$

where, as above, we have $+$ for fermions and $-$ for bosons.

We can also represent the average particle number as

$$N = \sum_s \langle n_s \rangle, \tag{7.22}$$

so we may identify

$$\langle n_s \rangle = \frac{1}{e^{\beta(\epsilon_s - \mu)} \pm 1}. \tag{7.23}$$

Although it may not immediately seem to be the case, the difference between the + and − sign in the denominator has important consequences for the quite different behaviour of fermions and bosons. If there are interactions between particles then we can't treat different energy levels as independent and the occupation numbers do not simplify to the forms in Eq. (7.23).

## 7.3.1  Fermi–Dirac Distribution

We have established in Eq. (7.23) that the occupation number for fermions in a state with energy $\epsilon_s$ is given by the Fermi–Dirac distribution:

$$\langle n_s \rangle = \frac{1}{e^{\beta(\epsilon_s - \mu)} + 1}. \tag{7.24}$$

The right-hand side of Eq. (7.24) is also known as the Fermi function, which is illustrated for several different temperatures in Fig. 7.2.

An important feature of the + sign in the denominator in Eq. (7.24) is that it ensures that $\langle n_s \rangle \leq 1$ for any choice of $\beta$, $\mu$ and $\epsilon_s$, in accordance with our expectation that we cannot have more than one fermion in a state. It is instructive to investigate various limits of the distribution. As $T \to 0$, there is a strong difference in the behaviour of $\langle n_s \rangle$, depending on whether $\epsilon_s$ is greater or less than the chemical potential $\mu$. If $\epsilon_s > \mu$, then $\langle n_s \rangle \to 0$ as $T \to 0$, whereas if $\epsilon_s < \mu$, $\langle n_s \rangle \to 1$, so that the zero-temperature limit of the Fermi function is a step function. The zero-temperature chemical potential is known as the **Fermi energy**, $\epsilon_F$.

To take the classical limit we need to have the average occupancy of each level be small compared to 1 (i.e. the dilute limit). In this case $e^{\beta(\epsilon_s - \mu)} \gg 1$, hence $e^{-\beta(\epsilon_s - \mu)} \ll 1$. Thus

$$\langle n_s \rangle = \frac{e^{-\beta(\epsilon_s - \mu)}}{1 + e^{-\beta(\epsilon_s - \mu)}} \simeq e^{-\beta(\epsilon_s - \mu)}, \tag{7.25}$$

which is the Maxwell–Boltzmann distribution, in agreement with our earlier arguments that this should be the classical limiting distribution for indistinguishable quantum particles.

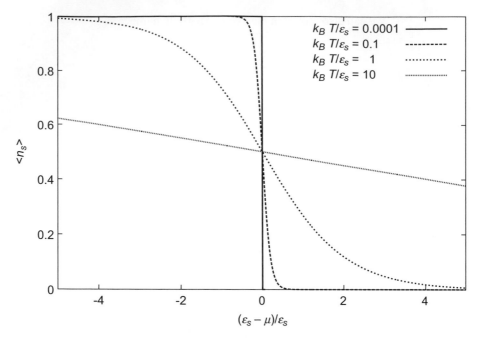

**Fig. 7.2**   Fermi–Dirac distribution as a function of energy over a wide range of temperatures.

### 7.3.2  Bose–Einstein Distribution

The average occupation number for bosons in a state with energy $\epsilon_s$ is given by the Bose–Einstein distribution, as we determined in Eq. (7.23):

$$\langle n_s \rangle = \frac{1}{e^{\beta(\epsilon_s - \mu)} - 1}. \tag{7.26}$$

The behaviour of this distribution is quite distinct from the Fermi–Dirac distribution. As $T \to 0$, if $\epsilon_s > \mu$, then $\langle n_s \rangle = 0$, however if $\epsilon_s < \mu$, we would get $\langle n_s \rangle = -1$, which is clearly nonsensical. This is resolved by the chemical potential behaving in such a way that there is never a state for which $\epsilon_s < \mu$, but $\mu \to \epsilon_0$, the lowest-energy eigenvalue, as $T \to 0$. If we were to allow $\mu = \epsilon_0$, then we would have the situation $\langle n_0 \rangle = \infty$ at $T = 0$.

We can take the classical limit of the Bose–Einstein distribution in a similar manner to the approach we used for fermions, and we find

$$\langle n_s \rangle = \frac{e^{-\beta(\epsilon_s - \mu)}}{1 - e^{-\beta(\epsilon_s - \mu)}} \simeq e^{-\beta(\epsilon_s - \mu)}, \tag{7.27}$$

showing that bosons also obey the Maxwell–Boltzmann distribution in the classical limit. The behaviours of the Bose–Einstein, Fermi–Dirac and Maxwell–Boltzmann distributions are compared in Fig. 7.3.

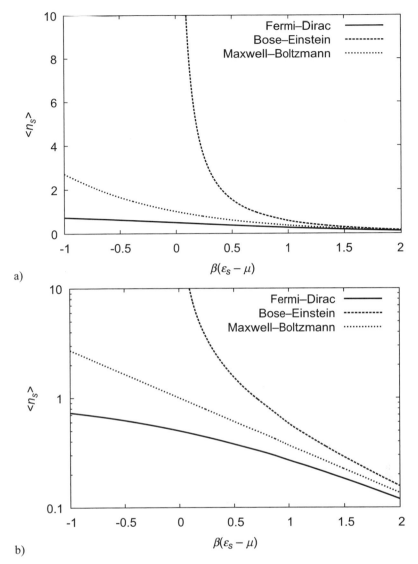

Fig. 7.3 Comparison of the Fermi–Dirac, Bose–Einstein and Maxwell–Boltzmann distributions at a temperature $k_B T = 0.1\epsilon_s$ on (a) a linear plot and (b) a log-linear plot.

## 7.4 Density of States and Thermal Averages

We discussed the density of states for the specific system of particles in a box in our derivation of the Maxwell–Boltzmann velocity distribution in Chapter 5. We will make extensive use of the density of states in calculating thermodynamic properties of quantum particles, and hence it is worth revisiting the topic.

We can calculate the density of states for free particles as follows. Consider free particles (i.e. there is no external potential and the particles do not interact) that are in a cubic box of side length $L$, subject to periodic boundary conditions. The Hamiltonian is

$$H = \frac{p^2}{2m} = -\frac{\hbar^2}{2m}\nabla^2, \tag{7.28}$$

and the allowed values of $x$, $y$ and $z$ vary between $-L/2$ and $L/2$. The single-particle eigenstates in this system can be written as

$$\psi = \left(\frac{1}{L}\right)^{\frac{3}{2}} e^{i\mathbf{k}\cdot\mathbf{r}}, \tag{7.29}$$

where $\mathbf{k} = \frac{2\pi}{L}(n_x, n_y, n_z)$ and $\mathbf{r} = (x, y, z)$, where $n_x$, $n_y$ and $n_z$ may take any integer value (i.e. negative as well as positive). In the space of wavevectors, the eigenstates form a regular grid, with a volume of $\left(\frac{2\pi}{L}\right)^3$ in $k$-space per eigenstate (in three dimensions). Equivalently, the density of plane waves in $k$-space is

$$\frac{1}{\left(\frac{2\pi}{L}\right)^3} = \frac{V}{8\pi^3}. \tag{7.30}$$

The energy eigenvalue associated with a state with wavevector $\mathbf{k}$ is

$$E = \langle\psi| H |\psi\rangle = \frac{\hbar^2 k^2}{2m} = \frac{2\pi^2\hbar^2}{mL^2}(n_x^2 + n_y^2 + n_z^2). \tag{7.31}$$

We can see that the surfaces of constant energy will be spherical in three-dimensional $k$-space. The total number of states within such a sphere can be calculated by adding up the numbers of states in concentric shells. The number of states in a thin shell of thickness $dk$ (as illustrated in Fig. 7.4 for two dimensions, but we will calculate for three dimensions) is

$$dN = \frac{g_s 4\pi k^2 dk}{\left(\frac{2\pi}{L}\right)^3}, \tag{7.32}$$

and noting that

$$k = \sqrt{\frac{2mE}{\hbar^2}} \implies k\, dk = \frac{m\, dE}{\hbar^2}, \tag{7.33}$$

we can rewrite Eq. (7.32) as

$$dN = \frac{g_s V}{2\pi^2}\sqrt{\frac{2mE}{\hbar^2}}\frac{m\, dE}{\hbar^2}$$

$$= \frac{g_s V}{4\pi^2}\left(\frac{2m}{\hbar^2}\right)^{\frac{3}{2}}\sqrt{E}\, dE. \tag{7.34}$$

Thus the density of states in three dimensions is

$$g(E) = \frac{dN}{dE} = \frac{g_s V}{4\pi^2}\left(\frac{2m}{\hbar^2}\right)^{\frac{3}{2}}\sqrt{E}. \tag{7.35}$$

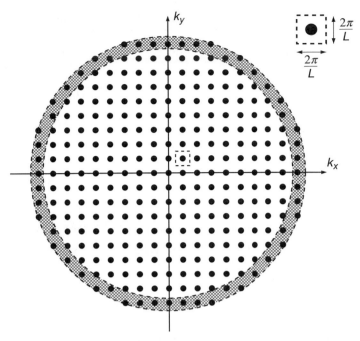

Fig. 7.4 States in *k*-space in two dimensions; surfaces of constant energy are dashed circles. The shaded area is an annulus of thickness *dk* that can be used to calculate *dN*. There is an area of $2\pi/L \times 2\pi/L$ associated with each state.

Unless we specify otherwise, we will assume going forward that particles are non-relativistic and in three dimensions, so that Eq. (7.35) is the relevant density of states. The same method we used to find the density of states in three dimensions can be used for one and two dimensions. It is easy to show that in two dimensions

$$dN = \frac{g_s 2\pi k \, dk}{\left(\frac{2\pi}{L}\right)^2}, \tag{7.36}$$

and in one dimension

$$dN = \frac{g_s 2dk}{\left(\frac{2\pi}{L}\right)}, \tag{7.37}$$

which can be used to find the density of states in energy by following similar steps to those we used to get from Eq. (7.32) to Eq. (7.35).

## 7.4.1  Thermal Averages Using the Density of States

In writing the probability of occupation of an energy level in terms of distribution functions that depend on the particle statistics, we have an alternative to taking derivatives of the grand canonical partition function and grand potential to calculate thermal averages of many quantities. For some function of energy $y(E)$, the average is

$$\langle y(E) \rangle = \sum_E y(E) p(E)$$

$$= \frac{\sum_E y(E) \langle n(E) \rangle}{\sum_E \langle n(E) \rangle}$$

$$= \frac{1}{N} \sum_E y(E) \langle n(E) \rangle, \tag{7.38}$$

where

$$p(E) = \frac{\langle n(E) \rangle}{\sum_E \langle n(E) \rangle} \tag{7.39}$$

is the probability that the system is in a state with energy $E$. Often we are dealing with systems in which the energy levels are closely spaced enough that we can treat the energy as a continuous variable, in which case we can replace the sum over discrete $E$ by an integral over continuous $E$. As we discussed above, we need to take into account the degeneracy of states with the same energy by using the density of states, $g(E)$, so

$$\sum_E \rightarrow \int dE \, g(E),$$

and we obtain

$$\langle y(E) \rangle = \frac{1}{N} \int_0^\infty dE \, g(E) \, y(E) f(E), \tag{7.40}$$

where $f(E) = \langle n(E) \rangle$ is the relevant distribution (Maxwell–Boltzmann, Fermi–Dirac or Bose–Einstein).

Note that $\langle y(E) \rangle$ is an average per particle. If we are interested in extensive quantities, such as the energy of a system, the relevant extensive quantity is $\langle Y(E) \rangle = N \langle y(E) \rangle$. For instance, the average energy of a system with $N$ particles is

$$U = \int_0^\infty dE \, g(E) \, E \, f(E), \tag{7.41}$$

and

$$N = \int_0^\infty dE \, g(E) \, f(E). \tag{7.42}$$

## 7.5 Summary

Quantum particles usually come in one of two varieties: fermions and bosons. There can be at most one fermion in a given energy level, whereas there is no such restriction on bosons. These differences in allowed occupations have implications for their statistical properties. The occupation of energy levels for each type of particle is described by the

**Fermi–Dirac distribution**

$$\langle n_s \rangle = \frac{1}{e^{\beta(\epsilon_s - \mu)} + 1} \qquad \text{(fermions)} \tag{7.43}$$

and the

**Bose–Einstein distribution**

$$\langle n_s \rangle = \frac{1}{e^{\beta(\epsilon_s - \mu)} - 1} \qquad \text{(bosons)}. \tag{7.44}$$

In the classical limit of low occupation per energy level, both the Fermi–Dirac and Bose–Einstein distributions simplify to the Maxwell–Boltzmann distribution

$$\langle n_s \rangle = e^{\beta(\mu - \epsilon_s)}, \tag{7.45}$$

which can be obtained for undistinguished particles, which are said to satisfy Maxwell–Boltzmann statistics.

# Problems

**7.1**   (a)  How many ways can $N$ bosons occupy $M$ orbitals?
         (b)  How many ways can $N$ fermions occupy $M$ orbitals?

**7.2**   Find the density of states $g(E)$ in one and two dimensions. Compare the energy dependence to the case of three dimensions that we found in Eq. (7.35).

**7.3**   The form of the canonical partition function for quantum systems is

$$Z = \text{Tr}\left[e^{-\beta H}\right],$$

where $H$ is the Hamiltonian. By evaluating the trace in a suitable basis, show that this expression reduces to the form we have already introduced:

$$Z = \sum_s e^{-\beta E_s}.$$

**7.4**   (a)  Use Eq. (7.11) for the grand potential of Maxwell–Boltzmann particles to calculate the chemical potential for a two-dimensional Maxwell–Boltzmann gas.
         (b)  Again using Eq. (7.11), calculate the entropy of a two-dimensional Maxwell–Boltzmann gas and use your answer to part (a) to eliminate the chemical potential in your expression. This is a two-dimensional analogue of the Sackur–Tetrode equation.

**7.5**   Show that the ideal gas law holds for an ultra-relativistic Maxwell–Boltzmann gas (i.e. a gas of particles for which the rest mass energy can be ignored and $E = |\mathbf{p}|c$).

**7.6**   (a)  Suppose we have a type of quantum particle for which the allowed occupancies of an energy level are 0, 1 and 2. If a system of such particles is non-interacting and connected to a thermal and diffusive reservoir, find the mean thermal occupancy of an energy level $\epsilon$.

(b) Now consider fermions. If there is an energy level that is doubly degenerate, i.e. two quantum states have the same energy $\epsilon$ (e.g. up- and down-spin states in zero magnetic field), find an expression for the mean thermal occupancy for energy $\epsilon$.

(c) Plot the number distributions you found in parts (a) and (b) and compare them. It may help to consider $\frac{\partial \langle n \rangle}{\partial \epsilon}\big|_{\epsilon=\mu}$.

**7.7**   An impurity atom in a semiconductor may bind zero, one (spin up or spin down) or two electrons. Assume that the binding energy for a single electron is $\epsilon$ (independent of spin), and that for two electrons the binding energy is $2\epsilon + U$, where $U$ reflects the Coulomb repulsion between the electrons. Find the average number of electrons bound to the impurity atom and compare this to a Fermi–Dirac distribution.

*Hint:* it may be useful to consider the limit $U = 0$.

# 8 Fermions

We have already seen in Chapter 7 that the Pauli exclusion principle gives rise to the Fermi–Dirac distribution, which is qualitatively distinct from the occupation number distribution for bosons. We found this distribution (Eq. (7.24)) to be

$$f_0(\epsilon) = \frac{1}{e^{\beta(\epsilon - \mu)} + 1}. \tag{8.1}$$

We first explore some of the thermodynamic properties that follow from this distribution for ideal and real Fermi gases and introduce the Fermi energy, a key quantity in fermionic systems. We then consider the Fermi sea in metals and electron degeneracy pressure in white dwarf stars as examples where Fermi–Dirac statistics have consequences for physics at macroscopic scales.

## 8.1 Chemical Potential for Fermions

In our expressions for both the Fermi–Dirac and Bose–Einstein distributions, the mean occupation for an energy level is expressed in terms of the chemical potential $\mu$. However, we have not actually specified expressions for the chemical potential for either fermions or bosons. In fermionic and bosonic systems the chemical potential needs to be determined by finding the value of $\mu$ that leads to the correct mean number of particles, i.e. in fermionic systems $\mu$ is determined so that

$$N = \sum_s \langle n_s \rangle = \sum_s \frac{1}{e^{\beta(\epsilon_s - \mu)} + 1}. \tag{8.2}$$

Recall that for a classical ideal gas (i.e. the Maxwell–Boltzmann distribution) we already have an expression for the chemical potential and do not need to follow this approach. Inverting Eq. (8.2) to determine $\mu$ at arbitrary temperature $T$ is non-trivial in general.

### 8.1.1 Zero Temperature: The Fermi Energy

At zero temperature, the Fermi function takes the form

$$f_0(\epsilon) = \theta(\mu - \epsilon), \tag{8.3}$$

where $\theta(x)$ is the Heaviside step function:

$$\theta(x) = \begin{cases} 1, & x > 0 \\ 0, & x < 0 \end{cases}. \tag{8.4}$$

The $T = 0$ form of the distribution allows us to calculate the chemical potential $\mu$ at $T = 0$ for free particles. We start by writing the particle number in the form of Eq. (7.42):

$$N = \frac{\partial \Phi}{\partial \mu} = \sum_s \langle n_s \rangle$$

$$= \int_0^\infty d\epsilon \, g(\epsilon) \, f_0(\epsilon). \tag{8.5}$$

We consider fermions in three dimensions with spin degeneracy $g_s$, hence we may use the form of the density of states found in Eq. (7.35) to write

$$N = \frac{g_s V}{4\pi^2} \left(\frac{2m}{\hbar^2}\right)^{\frac{3}{2}} \int_0^\infty d\epsilon \, \sqrt{\epsilon} \, f_0(\epsilon). \tag{8.6}$$

Our derivation to this point is valid for arbitrary $T$; if we assume that $T = 0$ this allows us to set the upper limit on the energy integral to $\mu$ and $f_0(\epsilon) = 1$ in the integrand, hence at zero temperature

$$N = \frac{g_s V}{4\pi^2} \left(\frac{2m}{\hbar^2}\right)^{\frac{3}{2}} \int_0^\mu d\epsilon \, \sqrt{\epsilon}$$

$$= \frac{g_s V}{6\pi^2} \left(\frac{2m\mu}{\hbar^2}\right)^{\frac{3}{2}}. \tag{8.7}$$

At zero temperature, the chemical potential is known as the Fermi energy $\epsilon_F$, which we can obtain by inverting the expression above:

$$\mu(T = 0) = \epsilon_F = \frac{\hbar^2}{2m} \left(\frac{6\pi^2 n}{g_s}\right)^{\frac{2}{3}}. \tag{8.8}$$

We can associate a temperature scale, the Fermi temperature $T_F = \epsilon_F / k_B$, with the Fermi energy. At temperatures $T \ll T_F$, the step-function-like nature of the Fermi–distribution is important, while at temperatures $T \gtrsim T_F$, the Fermi–Dirac distribution tends towards a Maxwell–Boltzmann distribution.

The zero-temperature limit that we have discussed above in which all states with energies lower than the Fermi energy are occupied and all states with higher energy are unoccupied is known as the *degenerate* limit and is illustrated in Fig. 8.1 for spin-1/2 fermions. It is straightforward to calculate thermodynamic properties of a Fermi gas in this limit, using a similar approach to the one we used above to calculate $N$. We will also be interested in the nearly degenerate limit $T \ll T_F$, for which the temperature is low enough that the Fermi–Dirac distribution has the appearance of a rounded step function. In this limit we can develop a perturbative expansion around the zero-temperature results known as the Sommerfeld expansion. We now introduce this approach using the chemical potential as an example.

### 8.1.2 Non-zero Temperature: Sommerfeld Expansion

To calculate the chemical potential at finite temperature, we need to return to the expression we obtained for $N$ that is valid at all temperatures, Eq. (8.5):

$$N = \int_0^\infty d\epsilon \, g(\epsilon) f_0(\epsilon). \tag{8.9}$$

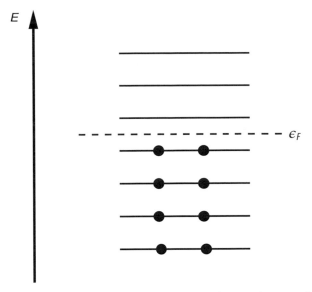

$E$

$\epsilon_F$

**Fig. 8.1** Spin-1/2 fermion system at zero temperature – there are two states at each energy due to spin degeneracy. All states with energy less than the Fermi energy $\epsilon_F$ are occupied and all states with energy greater than $\epsilon_F$ are unoccupied.

The calculation of finite-temperature corrections for many thermodynamic properties of an ideal Fermi gas involves similar integrals, and so we consider the more general integral

$$I = \int_0^\infty d\epsilon\, K(\epsilon) \frac{1}{e^{\beta(\epsilon-\mu)} + 1},$$
(8.10)

where $K(\epsilon)$ is some function that is smooth in comparison to the Fermi function in the vicinity of $\mu$. We will evaluate $I$ in the limit $T \to 0$, i.e. $\beta \to \infty$, and find corrections as a perturbative series in $T/T_F$. Write $K(\epsilon) = dG(\epsilon)/d\epsilon$ for an appropriate function $G(\epsilon)$, then we can integrate by parts and

$$I \simeq -G(0) + \int_0^\infty d\epsilon\, G(\epsilon) \left( -\frac{\partial f_0}{\partial \epsilon} \right),$$
(8.11)

provided $G$ does not have faster than polynomial growth in $\epsilon$. In the limit $\beta \to \infty$,

$$\left( -\frac{\partial f_0}{\partial \epsilon} \right) \to \delta(\epsilon - \mu),$$
(8.12)

so the derivative is very sharply peaked around $\epsilon = \mu$, in which case most of the contributions to the integral in Eq. (8.11) come from energies $\epsilon \sim \mu$. This suggests that we should be able to obtain an accurate approximation to $I$ by making an expansion of $G$ about $\epsilon = \mu$:

$$G(\epsilon) = G(\mu) + (\epsilon - \mu)G'(\mu) + \frac{1}{2!}(\epsilon - \mu)^2 G''(\mu) + \cdots,$$
(8.13)

and hence

$$I \simeq -G(0) + G(\mu) \int_0^\infty d\epsilon \left(-\frac{\partial f_0}{\partial \epsilon}\right) + G'(\mu) \int_0^\infty d\epsilon \, (\epsilon - \mu) \left(-\frac{\partial f_0}{\partial \epsilon}\right)$$
$$+ \frac{1}{2!} G''(\mu) \int_0^\infty d\epsilon \, (\epsilon - \mu)^2 \left(-\frac{\partial f_0}{\partial \epsilon}\right) + \cdots . \tag{8.14}$$

To evaluate the energy integrals, it is convenient to change variables to $x = \beta(\epsilon - \mu)$ and multiply numerator and denominator by $e^{-x}$, then the term in the expansion which is the coefficient of $G^{(r)}(\mu)$ becomes

$$\frac{1}{r!} \int_0^\infty d\epsilon \, (\epsilon - \mu)^r \, \frac{\beta e^{\beta(\epsilon - \mu)}}{(e^{\beta(\epsilon - \mu)} + 1)^2} = \frac{(k_B T)^r}{r!} \int_{-\mu/k_B T}^\infty dx \, \frac{x^r}{(1 + e^x)(1 + e^{-x})}$$
$$\simeq \frac{(k_B T)^r}{r!} \int_{-\infty}^\infty dx \, \frac{x^r}{(1 + e^x)(1 + e^{-x})}, \tag{8.15}$$

where the lower limit of the integral is taken to be $-\infty$ in the limit that $T \to 0$, since the contribution to the integral from the region with $-\infty < x < -\mu/k_B T$ will be exponentially small. Note that the integral vanishes whenever $r$ is odd, since the integrand is odd in that case, hence the lowest-order correction is of order $(k_B T)^2$:

$$\frac{1}{2!} G''(\mu) \int_0^\infty d\epsilon \, (\epsilon - \mu)^2 \left(-\frac{\partial f_0}{\partial \epsilon}\right) \simeq \frac{(k_B T)^2}{2} G''(\mu) \int_{-\infty}^\infty dx \, \frac{x^2}{(1 + e^x)(1 + e^{-x})}$$
$$= (k_B T)^2 G''(\mu) \int_0^\infty dx \, \frac{x^2}{(1 + e^x)(1 + e^{-x})}. \tag{8.16}$$

To perform the integral on the right-hand side of Eq. (8.16), note that

$$\frac{x^2}{(1 + e^x)(1 + e^{-x})} = \frac{x^2 e^x}{(1 + e^x)^2} = -x \frac{\partial}{\partial \alpha} \left[\frac{1}{1 + e^{\alpha x}}\right]\bigg|_{\alpha=1}, \tag{8.17}$$

hence

$$\int_0^\infty dx \, \frac{x^2}{(1 + e^x)(1 + e^{-x})} = -\frac{\partial}{\partial \alpha} \left[\int_0^\infty dx \, \frac{x e^{-\alpha x}}{1 + e^{-\alpha x}}\right]\bigg|_{\alpha=1}$$
$$= -\sum_{n=0}^\infty (-1)^n \frac{\partial}{\partial \alpha} \left[\int_0^\infty dx \, x \, e^{-\alpha(n+1)x}\right]\bigg|_{\alpha=1}$$
$$= -\sum_{n=0}^\infty \frac{(-1)^n}{(n+1)^2} \left\{\frac{\partial}{\partial \alpha}\left[\frac{1}{\alpha^2}\right]\right\}\bigg|_{\alpha=1} \int_0^\infty dy \, y \, e^{-y}, \tag{8.18}$$

where we rewrote

$$\frac{1}{1 + e^{-\alpha x}} = \sum_{n=0}^\infty (-1)^n e^{-n\alpha x} \tag{8.19}$$

and changed variables to $y = \alpha(n + 1)x$. After performing the integral over $y$, Eq. (8.18) can be simplified to

$$\int_0^\infty dx \frac{x^2}{(1 + e^x)(1 + e^{-x})} = 2\Gamma(2) \sum_{n=0}^\infty \frac{(-1)^n}{(n + 1)^2}, \tag{8.20}$$

where $\Gamma(2)$ is an example of the gamma function (Eq. (A.6)), which is discussed in detail in Section A.2. Noting that odd terms in the sum come with a $+$ sign and even terms come with a $-$ sign, we can separate the sums over odd and even $n$ on the right-hand side of Eq. (8.20) to get

$$\int_0^\infty dx \frac{x^2}{(1 + e^x)(1 + e^{-x})} = 2 \left\{ \sum_{n=0}^\infty \frac{1}{(2n + 1)^2} - \sum_{n=1}^\infty \frac{1}{(2n)^2} \right\}$$

$$= 2 \left\{ \sum_{n=1}^\infty \frac{1}{n^2} - 2 \sum_{n=1}^\infty \frac{1}{(2n)^2} \right\}$$

$$= \sum_{n=1}^\infty \frac{1}{n^2}$$

$$= \frac{\pi^2}{6}, \tag{8.21}$$

where we made use of the identity

$$\sum_{n=1}^\infty \frac{1}{n^2} = \frac{\pi^2}{6}. \tag{8.22}$$

Thus

$$I \simeq -G(0) + G(\mu) + \frac{\pi^2}{6}(k_B T)^2 G''(\mu) + \cdots, \tag{8.23}$$

or in terms of the function $K(\epsilon)$:

$$I \simeq \int_0^\mu d\epsilon \, K(\epsilon) + \frac{\pi^2}{6}(k_B T)^2 K'(\mu) + O(T^4) + \cdots. \tag{8.24}$$

This result is the Sommerfeld expansion for a quantity $K(\epsilon)$. If the temperature is zero then $\mu = \epsilon_F$, but for non-zero temperature $\mu \neq \epsilon_F$, so in order to make use of the expansion we must first find the temperature dependence of the chemical potential.

### 8.1.3  Temperature Dependence of the Chemical Potential

We can apply Eq. (8.24) to the particular case of the particle number $N$, in which case

$$K(\epsilon) = g(\epsilon) = \frac{g_s V}{4\pi^2} \left( \frac{2m}{\hbar^2} \right)^{\frac{3}{2}} \epsilon^{\frac{1}{2}} \tag{8.25}$$

is the density of states in three dimensions. Hence

$$N \simeq \frac{g_s V}{4\pi^2} \left( \frac{2m}{\hbar^2} \right)^{\frac{3}{2}} \left[ \frac{2}{3} \mu^{\frac{3}{2}} + \frac{\pi^2}{6}(k_B T)^2 \frac{1}{2\sqrt{\mu}} + \cdots \right]. \tag{8.26}$$

To find $\mu$ as a function of $T$ we must solve Eq. (8.26) for $\mu$. We can note that if we are solving to lowest order in $T$, then in the second term in square brackets in Eq. (8.26) we can replace $\mu$ by $\mu(T = 0) = \epsilon_F$ without any loss of accuracy. Then

$$\mu^{\frac{3}{2}} \simeq \epsilon_F^{\frac{3}{2}} - \frac{3}{2}\frac{\pi^2}{6}\frac{(k_BT)^2}{2\sqrt{\epsilon_F}}, \tag{8.27}$$

where we used Eq. (8.8) to simplify Eq. (8.26). When we take the two-thirds root of Eq. (8.27) we get (making use of $(1 + x)^\alpha \simeq 1 + \alpha x$ for small $x$)

$$\mu \simeq \epsilon_F\left[1 - \frac{\pi^2}{12}\frac{T^2}{T_F^2}\right]. \tag{8.28}$$

The finite-temperature chemical potential $\mu(T)$ is often referred to as the Fermi level (which is equal to the Fermi energy $\epsilon_F$ when $T = 0$).

## 8.2 Thermodynamic Properties of a Fermi Gas

Now that we know the temperature dependence of the chemical potential, we are in a position to use it in Eq. (8.24) to calculate the finite-temperature thermodynamic properties of a Fermi gas.

### 8.2.1 Energy and Heat Capacity

Making use of Eq. (7.41) we can write down the general expression for the energy of a Fermi gas at arbitrary temperature as

$$U = \int_0^\infty d\epsilon\, g(\epsilon)\,\epsilon\, f_0(\epsilon), \tag{8.29}$$

and in the zero-temperature limit we can simplify the integral in a similar manner to our calculation of the Fermi energy:

$$U = \frac{g_sV}{4\pi^2}\left(\frac{2m}{\hbar^2}\right)^{\frac{3}{2}}\int_0^{\epsilon_F} d\epsilon\,\epsilon^{\frac{3}{2}} = \frac{3}{5}N\epsilon_F. \tag{8.30}$$

Note that the average energy per particle is not equal to the Fermi energy – the Fermi energy is the maximum energy a particle can have in the degenerate limit, not the average.

To obtain finite-temperature corrections to the energy, we can refer to the Sommerfeld expansion with $K(\epsilon) = \epsilon g(\epsilon)$, and so (to lowest order in $T/T_F$)

$$\begin{aligned}
U &\simeq \frac{2}{5}\frac{g_sV}{4\pi^2}\left(\frac{2m}{\hbar^2}\right)^{\frac{3}{2}}\mu^{\frac{5}{2}} + \frac{g_sV}{4\pi^2}\left(\frac{2m}{\hbar^2}\right)^{\frac{3}{2}}\frac{3}{2}\mu^{\frac{1}{2}}\frac{\pi^2}{6}(k_BT)^2 \\
&\simeq \frac{3}{5}N\epsilon_F\left(1 - \frac{5\pi^2}{24}\frac{T^2}{T_F^2}\right) + \frac{3\pi^2}{8}\frac{T^2}{T_F^2}N\epsilon_F \\
&= \frac{3}{5}N\epsilon_F + \frac{\pi^2}{4}\frac{T^2}{T_F^2}N\epsilon_F,
\end{aligned} \tag{8.31}$$

where we used $(1 + x)^\alpha \sim 1 + \alpha x$ when substituting Eq. (8.28) for $\mu$ in the second line and only kept terms to order $(T/T_F)^2$. We can use Eq. (8.31) for the energy to infer that the heat capacity at low temperatures for a Fermi gas is

$$C_V = \left.\frac{\partial U}{\partial T}\right|_V = \frac{\pi^2}{2} N k_B \frac{T}{T_F}. \tag{8.32}$$

Linearity of the heat capacity with temperature at low temperatures is a signature of a system of non-interacting fermions, and the vanishing of the heat capacity as $T \to 0$ reflects the fact that the number of available states in which to deposit energy at very low temperatures vanishes as $T \to 0$, since almost all states with energies below the Fermi energy are occupied, and very few states with energies above the Fermi energy are occupied.

### 8.2.2 Pressure of a Fermi Gas

To calculate the pressure of a Fermi gas, start from the identity found in Eq. (6.42):

$$P = \left.\frac{\partial \Phi}{\partial V}\right|_{\mu,T}. \tag{8.33}$$

To determine the right-hand side of the equation we can take the grand potential for fermions and integrate by parts, so we have

$$\Phi = k_B T \sum_s \ln\left[1 + e^{\beta(\mu-\epsilon)}\right]$$

$$= k_B T \int_0^\infty d\epsilon\, g(\epsilon) \ln\left[1 + e^{\beta(\mu-\epsilon)}\right]$$

$$= k_B T \frac{g_s V}{4\pi^2} \left(\frac{2m}{\hbar^2}\right)^{\frac{3}{2}} \int_0^\infty d\epsilon\, \sqrt{\epsilon} \ln\left[1 + e^{\beta(\mu-\epsilon)}\right]$$

$$= k_B T \frac{g_s V}{4\pi^2} \left(\frac{2m}{\hbar^2}\right)^{\frac{3}{2}} \left\{\left[\frac{2}{3}\epsilon^{\frac{3}{2}} \ln\left(1 + e^{\beta(\mu-\epsilon)}\right)\right]_0^\infty + \frac{2\beta}{3} \int_0^\infty d\epsilon\, \epsilon^{\frac{3}{2}} \frac{1}{e^{\beta(\epsilon-\mu)} + 1}\right\}$$

$$= \frac{2}{3} \int_0^\infty d\epsilon\, g(\epsilon)\, \epsilon\, f_0(\epsilon)$$

$$= \frac{2}{3} U. \tag{8.34}$$

We note that $g(\epsilon) \propto V$, hence a derivative with respect to $V$ will give

$$P = \frac{\partial \Phi}{\partial V} = \frac{2}{3}\frac{U}{V}. \tag{8.35}$$

Note that this pressure is non-zero even at zero temperature and in the degenerate limit is known as degeneracy pressure. This is a consequence of the Pauli exclusion principle – in a fermionic system the kinetic energy is non-zero even at $T = 0$ due to contributions from levels with energies all the way up to the Fermi energy. We can compare the expression for $P$ to the expression for the grand potential, Eq. (8.34), and we see that this leads to the expression

$$\Phi = PV \tag{8.36}$$

for a three-dimensional Fermi gas.

### 8.2.3 Entropy of a Fermi Gas

To calculate the entropy of a Fermi gas, start from

$$S = \left.\frac{\partial \Phi}{\partial T}\right|_{\mu,V}$$
$$= \sum_s \left\{ k_B \ln\left(1 + e^{\beta(\mu-\epsilon_s)}\right) + \frac{k_B\beta(\epsilon_s - \mu)}{e^{\beta(\epsilon_s-\mu)} + 1} \right\}, \tag{8.37}$$

and then use

$$1 + e^{\beta(\mu-\epsilon_s)} = \frac{1}{1 - \langle n_s \rangle} \tag{8.38}$$

and

$$\beta(\epsilon_s - \mu) = \ln\left(\frac{\langle n_s \rangle}{1 - \langle n_s \rangle}\right), \tag{8.39}$$

to see that

$$S = \sum_s \left\{ -k_B \ln\left(1 - \langle n_s \rangle\right) - k_B \langle n_s \rangle \left[\ln \langle n_s \rangle - \ln\left(1 - \langle n_s \rangle\right)\right] \right\}$$
$$= -k_B \sum_s \left\{ \langle n_s \rangle \ln \langle n_s \rangle + \left(1 - \langle n_s \rangle\right) \ln\left(1 - \langle n_s \rangle\right) \right\}. \tag{8.40}$$

We know that at low temperatures, $\langle n_s \rangle$ is either very close to 0 or 1 for most energy levels, in which case most of the terms in Eq. (8.40) are negligible, so the entropy is dominated by the levels in an energy window of width $\sim k_B T$ around the chemical potential for which $\langle n_s \rangle$ differs significantly from zero.

If we denote the probability that energy level $s$ is unoccupied as $P_s^0$ and the probability that energy level $s$ is occupied as $P_s^1$, then we have the equations

$$P_s^0 + P_s^1 = 1 \tag{8.41}$$

and

$$(0 \times P_s^0) + (1 \times P_s^1) = \langle n_s \rangle. \tag{8.42}$$

Then we can read off that $P_s^1 = \langle n_s \rangle$ and $P_s^0 = 1 - \langle n_s \rangle$, which allows us to write

$$S = -k_B \sum_s \left[ P_s^0 \ln P_s^0 + P_s^1 \ln P_s^1 \right]. \tag{8.43}$$

This is in the form of the Gibbs entropy, where for each energy level $s$ we get a contribution from the probability that the energy level is either occupied or unoccupied. This allows us to interpret the entropy of a Fermi gas as the sum of our ignorance about the occupation of each of the energy levels $s$.

## 8.2.4 Number Fluctuations

A question we have not addressed up to this point is that of fluctuations in a non-interacting fermion system. We can calculate the fluctuations in a single level $s$ using a similar method to the one we used to rewrite the entropy:

$$\left\langle n_s^2 \right\rangle = (0)^2 \times P_s^0 + (1)^2 \times P_s^1 = \langle n_s \rangle, \tag{8.44}$$

so the relative rms fluctuations are

$$\frac{\left\langle (\Delta n_s)^2 \right\rangle}{\langle n_s \rangle^2} = \frac{\left\langle n_s^2 \right\rangle - \langle n_s \rangle^2}{\langle n_s \rangle^2} = \frac{1 - \langle n_s \rangle}{\langle n_s \rangle}. \tag{8.45}$$

This expression is not necessarily small, since $\langle n_s \rangle$ can take any value between 0 and 1, and so the rms fluctuations for an individual level take any value between 0 and $\infty$. Specifically, the rms fluctuations go to zero as $\langle n_s \rangle \to 1$ and the fluctuations go to infinity as $\langle n_s \rangle \to 0$. The fluctuations in a single level may be large, but if we focus on the total number of particles in the system

$$N = \langle N \rangle = \left\langle \sum_s n_s \right\rangle = \sum_s \langle n_s \rangle \tag{8.46}$$

and

$$\langle N^2 \rangle = \left\langle \left( \sum_s n_s \right) \left( \sum_r n_r \right) \right\rangle = \sum_s \sum_r \langle n_s n_r \rangle, \tag{8.47}$$

then the rms fluctuations in the total number of particles in the system are

$$\frac{\left\langle (\Delta N)^2 \right\rangle}{N^2} = \frac{1}{N^2} \left[ \sum_s \sum_r \langle n_s n_r \rangle - \sum_s \sum_r \langle n_s \rangle \langle n_r \rangle \right]. \tag{8.48}$$

We can separate the terms for which $s = r$ from those for which $s \neq r$ to get

$$\frac{\left\langle (\Delta N)^2 \right\rangle}{N^2} = \frac{1}{N^2} \sum_s \left[ \langle n_s^2 \rangle - \langle n_s \rangle^2 \right] + \frac{1}{N^2} \sum_s \sum_{r \neq s} [\langle n_s n_r \rangle - \langle n_s \rangle \langle n_r \rangle]. \tag{8.49}$$

The particles are non-interacting, so the occupancies of levels with different $r$ and $s$ are independent and we must have that $\langle n_s n_r \rangle = \langle n_s \rangle \langle n_r \rangle$, so the second sum in the expression above vanishes, and we are left with

$$\frac{\left\langle (\Delta N)^2 \right\rangle}{N^2} = \frac{1}{N^2} \sum_s \langle n_s \rangle (1 - \langle n_s \rangle) = \frac{1}{N} - \frac{1}{N^2} \sum_s \langle n_s \rangle^2, \tag{8.50}$$

and since $0 \leq \langle n_s \rangle \leq 1$, all the terms in the second sum also lie between 0 and 1, so $\sum_s \langle n_s \rangle^2 < \sum_s \langle n_s \rangle = N$, which implies

$$\frac{\left\langle (\Delta N)^2 \right\rangle}{N^2} \sim \frac{1}{N}. \tag{8.51}$$

Thus, relative fluctuations of the total particle number become negligible in the thermodynamic limit.

### 8.2.5 Another View of Temperature Dependence of Thermodynamic Properties

For temperatures $T \ll T_F$, the number of occupied states above the Fermi energy and the number of unoccupied states below the Fermi energy are approximately the same and so corrections to the zero-temperature results are small, as we have seen in the Sommerfeld expansion. An alternative, and more physical way to understand the finite-temperature correction to the zero-temperature average for some quantity $Q(\epsilon)$ can be obtained by writing the average in the following way:

$$
\begin{aligned}
\langle Q \rangle &= \int_0^\infty d\epsilon\, g(\epsilon)\, Q(\epsilon)\, f_0(\epsilon) \\
&= \int_0^{\epsilon_F} d\epsilon\, g(\epsilon)\, Q(\epsilon)\, f_0(\epsilon) + \int_{\epsilon_F}^\infty d\epsilon\, g(\epsilon)\, Q(\epsilon)\, f_0(\epsilon) \\
&= \langle Q \rangle_{T=0} + \int_{\epsilon_F}^\infty d\epsilon\, g(\epsilon)\, Q(\epsilon)\, f_0(\epsilon) - \int_0^{\epsilon_F} d\epsilon\, g(\epsilon)\, Q(\epsilon)\, (1 - f_0(\epsilon)),
\end{aligned}
\qquad (8.52)
$$

where

$$
\langle Q \rangle_{T=0} = \int_0^{\epsilon_F} d\epsilon\, g(\epsilon)\, Q(\epsilon)\, f_0(\epsilon). \qquad (8.53)
$$

This rearrangement of terms makes it clear that the deviations from $T = 0$ behaviour at finite temperature can be grouped into contributions of excitations with energies higher than the Fermi energy (i.e. energies $\epsilon$ where $f_0(\epsilon)$ becomes non-zero) and excitations with energies less than the Fermi energy (holes, which have occupation $1 - f_0(\epsilon) \neq 0$). This idea is illustrated in Fig. 8.2.

# 8.3 Applications

With the exception of some systems of trapped ultracold fermionic atoms, most fermionic systems do not approach the ideal Fermi gas limit. However, concepts such as the Fermi energy and fermion degeneracy pressure also have application in real systems. We illustrate this in the context of the Fermi surface in metals and degeneracy pressure in compact stars such as white dwarfs.

### 8.3.1 Metals and the Fermi Sea

In metals, the interactions between electrons are not small, however, in many metals a description of electronic properties in terms of non-interacting electrons is quite accurate at low temperatures $T \ll T_F \sim 10^4$ K. For atoms in metals, the nucleus and the inner core electrons are quite tightly bound, whereas the outer valence electrons can be considered as "nearly free" and moving in a periodic potential from the positively charged ions. This simplistic picture has considerable predictive power in many cases. If we look in momentum space at free electrons, then the Fermi energy defines a constant-energy surface

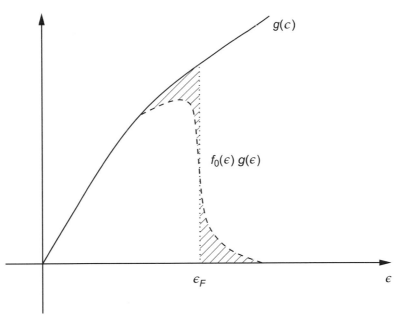

$g(\epsilon)$

$f_0(\epsilon)\, g(\epsilon)$

$\epsilon_F$

$\epsilon$

**Fig. 8.2**    Density of states in three dimensions, multiplied by the Fermi function at low temperature. States with energy $\epsilon < \epsilon_F$ that would be filled at zero temperature, but are unoccupied, and states with energy $\epsilon > \epsilon_F$ that would be empty at zero temperature, but are occupied at finite temperature, are shaded.

known as the Fermi surface. For free electrons, this is a sphere, with radius given by the Fermi wavevector

$$k_F = \frac{\sqrt{2m\epsilon_F}}{\hbar}. \tag{8.54}$$

In materials the Fermi surface is generally not a sphere and has a more complicated shape, reflecting the symmetry of the crystal lattice. As we have seen above, many physical quantities in Fermi systems are dominated by states near the Fermi energy, hence a clear understanding of the Fermi surface can lead to considerable insight into the properties of metals. The first determination of a Fermi surface of a metal was that of copper, using the anomalous skin effect, by Pippard in 1957, as illustrated in Fig. 8.3. There are now a variety of experimental and theoretical techniques that allow the determination of Fermi surfaces in metals. A short list includes the de Haas van Alphen effect, Shubnikov de Haas oscillations, angle-resolved photoemission spectroscopy (ARPES) and angle-dependent magnetoresistance oscillations (AMRO). An example of a more complex Fermi surface than copper, obtained for $Sr_2RuO_4$ using the de Haas van Alphen effect, is shown in Fig. 8.4. Even with the richness possible in Fermi surfaces, a lowest-order approximation of a spherical surface, which is much easier to calculate with than a more complicated geometry, can often give a qualitatively correct picture in real materials.

An example where the picture of nearly free electrons is effective is the heat capacity. The total low-temperature heat capacity in a metal is often well approximated by the form

$$C_V = \gamma T + \alpha T^3, \tag{8.55}$$

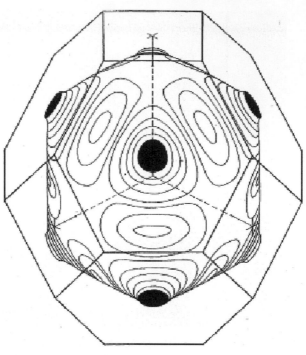

**Fig. 8.3**  Fermi surface of Cu as determined by A. B. Pippard (Pippard, 1957) using the anomalous skin effect. Reproduced from Pippard, A. B. 1957. An experimental determination of the Fermi surface in copper. *Phil. Trans. R. Soc. Lond., Ser. A*, **250**, 325–357

and for non-interacting fermions we calculated earlier that

$$\gamma = \frac{\pi^2}{2} \frac{N k_B}{T_F}. \tag{8.56}$$

Interactions between fermions lead to a modified value of $\gamma$ that is "renormalized" from the non-interacting value. The $T^3$ term in Eq. (8.55) arises from phonons (which we will discuss in Chapter 9). At room temperature the phonon contribution is much larger than the electronic contribution, but at low enough temperatures the electronic contribution is dominant.

Even though there are strong Coulomb interactions between electrons in most metals, the form of many of the low-temperature properties is the same as if the electrons did not interact, but with coefficients (e.g. $\gamma$) that are modified from their non-interacting values. The fact that a description of the properties of metals in terms of non-interacting fermions works so well, despite strong interactions between the electrons, is quite non-trivial. Lev Landau explained how this could arise by considering the thought experiment in which interactions are turned on slowly in a non-interacting fermion system. In this picture, the ground state of the non-interacting fermion system (i.e. all orbitals with energies less than $\epsilon_F$ are occupied) evolves to become the ground state of the interacting system. The excited states of the non-interacting system, in which fermions occupy orbitals with energies above the Fermi energy and leave behind holes in the occupied orbitals, evolve to become the

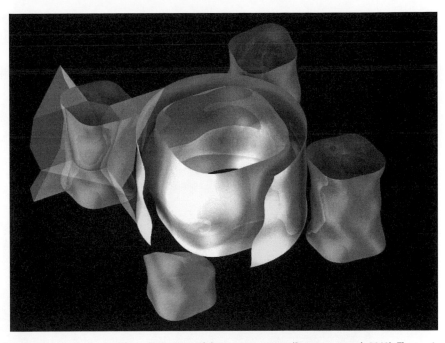

**Fig. 8.4**  Fermi surface of $Sr_2RuO_4$ obtained from de Haas van Alphen measurements (Bergemann *et al.*, 2003). The *c*-axis corrugation has been exaggerated by a factor of 15 for clarity. Reproduced from Bergemann, C., Mackenzie, A. P., Julian, S. R., Forsythe, D., and Ohmichi, E. 2003. Quasi-two-dimensional Fermi liquid properties of the unconventional superconductor $Sr_2RuO_4$. *Adv. Phys.*, **52**, 639–725. With permission of Taylor & Francis Ltd. (www.tandfonline.com)

excited states of the interacting system. These excited states, known as quasiparticles, have the same quantum numbers (spin, charge and momentum) as in the non-interacting system but are not completely stable – their lifetime gets shorter the further they are from the Fermi energy. Properties such as the mass and magnetic moment of the fermions can be quite strongly modified (known as renormalized) by interactions. These renormalized properties can be measured: the coefficient $\gamma$ that arises in the heat capacity at low temperatures is proportional to the renormalized fermion mass $m^*$, which can be found to be quite different from that of an electron. The full scope of how the properties of a non-interacting Fermi gas are modified are set out in what is known as Fermi liquid theory, the details of which are beyond the scope of our discussion here.

### 8.3.2  White Dwarf Stars

The first white dwarf star to be discovered was Eridani B (which is at a distance of approximately 17 light years from Earth) by William Herschel in 1783. The second was inferred in 1844 by Frederick Bessel, who noted that Sirius shows periodic variations in its position in the sky and suggested that it had an unseen companion. This companion, Sirius B, was discovered by Alvan Clark in 1862. In the early twentieth century, astronomers realized that white dwarf stars were much more dense than other stellar objects known at the time and had lower luminosity than other objects with similar mass. That white

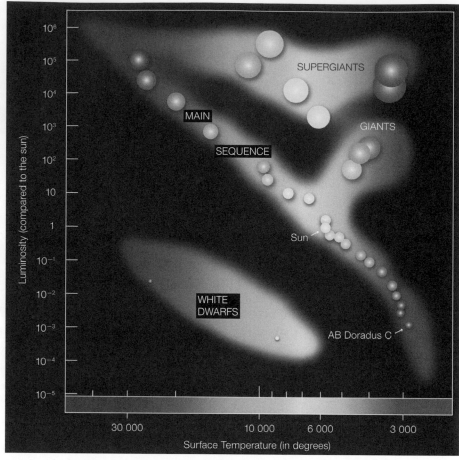

**Fig. 8.5** Hertzsprung–Russell diagram illustrating different populations of stars. Note that white dwarf stars occupy a distinct position in the diagram separate from main sequence stars such as the Sun. Reproduced with permission of the European Southern Observatory (ESO).

dwarf stars constitute a different stellar population than regular stars like the Sun can be seen on a Hertzsprung–Russell diagram which plots luminosity against temperature, as illustrated in Fig. 8.5. The masses of white dwarf stars are on the order of that of the Sun, $M_\odot \simeq 2 \times 10^{30}$ kg, but with a radius of order that of the Earth, $\sim 10^7$ m. The density of the Sun is of order 1000 kg m$^{-3}$ (i.e. roughly the same density as water), whereas white dwarf stars can have densities of order $10^8$–$10^9$ kg m$^{-3}$. When the density is so high, electrons are no longer bound to individual nuclei and form a degenerate electron gas. Even though white dwarfs can be quite hot, $\sim 10^7$ K, the Fermi temperature for electrons in such a system is $\sim 10^9$ K, so the Fermi gas of electrons is nearly degenerate and can actually be considered to be in the low-temperature limit! The pressure exerted by degenerate electrons in these stars is sufficient to prevent their collapse from their self-gravity.

Stars are massive hot balls of gas that maintain their size through an equilibrium between an outward pressure that balances the tendency to collapse under the gravitational attraction of the matter from which they are composed. In regular main sequence stars, the pressure has a thermal origin, from energy released by nuclear fusion. Eventually the fuel for nuclear reactions is used up so there is no pressure to oppose gravity, and the star collapses in either a nova or a supernova explosion depending on its mass, leaving a remnant. Both white dwarf stars and neutron stars are remnant stars supported by degeneracy pressure, which does not require an ongoing source of energy (as opposed to thermal pressure). Stars with a mass less than about eight times the mass of the Sun ($8M_\odot$) are believed to leave a white dwarf as a remnant, and more massive stars explode in a supernova before leaving a neutron star remnant. Supermassive stars (with masses tens of times larger than $M_\odot$) may lead to remnants where even degeneracy pressure is insufficient to prevent gravitational collapse, eventually leading to a black hole.

We will not concern ourselves with the dynamics of how a white dwarf forms, but will examine the resulting equilibrium situation. To treat the force balance inside a white dwarf quantitatively, let $\psi(r)$ be the gravitational potential of a star (which we will assume to be spherically symmetric) and let $\rho(r)$ be its mass density, where $r$ is the radial co-ordinate. From Newton's second law

$$F = ma = -\frac{GmM(r)}{r^2} = -m\frac{d\psi}{dr}, \tag{8.57}$$

where $m$ is a test mass at radius $r$ and $M(r)$ is the mass contained inside radius $r$. For a spherically symmetric star with density $\rho(r)$ that depends on position, the mass inside a spherical shell of radius $r$ and thickness $dr$ is

$$dM = 4\pi\rho(r)r^2, \tag{8.58}$$

and hence the total mass inside radius $r$ is

$$M(r) = \int_0^r dr'\, 4\pi r'^2\, \rho(r'), \tag{8.59}$$

with the boundary condition that $M(R) = M$, where $R$ is the radius of the star, and hence

$$r^2\frac{d\psi}{dr} = GM(r) = G\int_0^r dr'\, 4\pi r'^2\, \rho(r'). \tag{8.60}$$

Thus

$$\frac{1}{r^2}\frac{d}{dr}\left(r^2\frac{d\psi}{dr}\right) = 4\pi G\rho(r), \tag{8.61}$$

which relates the gravitational potential to the density profile.

The pressure gradient can also be related to the gravitational force by considering a spherical shell of radius $r$ and thickness $dr$ (Fig. 8.6). The outward force from the pressure on the shell is $dF_P = 4\pi r^2 dP$, where $dP$ is the pressure difference across the shell and $4\pi r^2$ is its surface area. The inward gravitational force on the shell is

$$dF_G = -\frac{GM(r)dm(r)}{r^2} = -\frac{GM(r)4\pi r^2\rho(r)}{r^2}, \tag{8.62}$$

where the mass in the shell is $dm(r) = 4\pi r^2\rho(r)dr$.

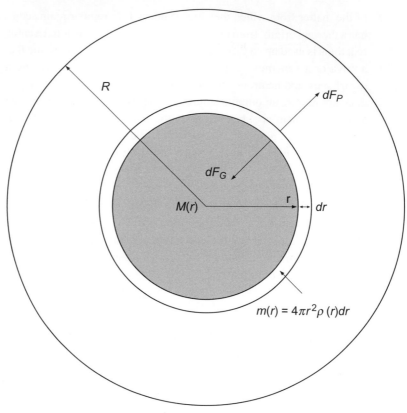

**Fig. 8.6**  Balance of the pressure force $dF_P$ and the gravitational force $dF_G$ for a shell of radius $r$, thickness $dr$ and density $\rho(r)$ in a stellar interior for a star of radius $R$.

When $dF_P$ and $dF_G$ are in equilibrium, then

$$\frac{dP(r)}{dr} = -\frac{GM(r)\rho(r)}{r^2}. \tag{8.63}$$

The boundary conditions on the pressure are that $P(R) = 0$ and $dP/dr = 0$ at $r = 0$. These equations also apply to a white dwarf star, but it is simpler to impose the force balance condition using a different argument. We determined the pressure in terms of the Fermi energy above (recall this only holds at $T = 0$) and we can relate $P(r)$ to the local Fermi energy $\epsilon_F(r)$. At $T = 0$ in the absence of a field, the Fermi energy is equal to the chemical potential, however, we need to recall that the chemical potential is the energy needed to add a particle to the system. To add a particle at radius $r$ we must also give it the gravitational potential energy required by a particle at that radius, which is $m\psi(r)$. For each electron we add, there must be a proton or another ion associated with it so that the star is electrically neutral, and so the mean mass per electron is $\zeta_e m_H$, where $m_H$ is the mass of a hydrogen atom.[1] In equilibrium, the chemical potential must be equal everywhere, hence

[1]  $\zeta_e$ will depend on the chemical composition of the star: for fully ionized $^4$He, $^{12}$C or $^{16}$O, each of which have equal numbers of protons and neutrons, $\zeta_e = 2$.

$$\mu = \epsilon_F(r) + \zeta_e m_H \psi(r),  \tag{8.64}$$

where $\psi(r)$ is determined by the mass density

$$\rho(r) = \zeta_e m_H n_e(r),  \tag{8.65}$$

and $n_e(r)$ is the local number density. This allows us to express $\psi(r)$ in terms of $\epsilon_F(r)$. In turn, the Fermi energy can be related back to the number density as

$$\epsilon_F(r) = \frac{\hbar^2}{2m_e}\left[3\pi^2 n_e(r)\right]^{\frac{2}{3}},  \tag{8.66}$$

for non-relativistic electrons, taking into account spin degeneracy $g_s = 2$. In the interior of white dwarf stars, electrons can have sufficient kinetic energy that it is not necessarily appropriate to assume that they are non-relativistic. In the ultra-relativistic limit their energy is large compared to the electron rest mass energy, so that $\epsilon \simeq pc$, in which case the local Fermi energy is related to the local number density by

$$\epsilon_F(r) = \hbar c \left[3\pi^2 n_e(r)\right]^{\frac{1}{3}}.  \tag{8.67}$$

If we introduce the dimensionless variable $\xi = r/R$ and write $\epsilon_F(r) = \epsilon_c f(\xi)$, then we can simplify the equation for the gravitational potential, Eq. (8.61), in either the non-relativistic or ultra-relativistic cases to

$$\frac{1}{\xi^2}\frac{d}{d\xi}\left[\xi^2 \frac{df(\xi)}{d\xi}\right] = \begin{cases} -[f(\xi)]^{\frac{3}{2}}, & \text{non-relativisitic case} \\ \\ -[f(\xi)]^3, & \text{ultra-relativistic case} \end{cases}.  \tag{8.68}$$

In order to do this, we use Eq. (8.64) to relate $d\psi/dr$ to $\epsilon_F(r)$ and rewrite $\rho(r)$ in Eq. (8.61) in terms of $\epsilon_F(r)$ by using either Eq. (8.66) or Eq. (8.67) in the non-relativistic and ultra-relativistic cases, respectively. After doing this, the energy $\epsilon_c$ in the two cases can be read off as

$$\epsilon_c = \begin{cases} \dfrac{3^2(2\pi\hbar)^6}{2^{13}\pi^4\zeta_e^4 m_H^4 m_e^3 G^2 R^4}, & \text{non-relativisitic case} \\ \\ \left[\dfrac{3(2\pi\hbar)^3 c^3}{32\pi^2\zeta_e^2 m_H^2 G R^2}\right]^{\frac{1}{2}}, & \text{ultra-relativistic case} \end{cases}.  \tag{8.69}$$

The equation for $f$ needs to be solved numerically, with boundary conditions $f'(0) = 0$ and $f(1) = 0$ that come from the boundary conditions on the pressure.

The mass of a white dwarf may be estimated from the central density of the star multiplied by the volume up to factors of order unity. The central density is $\zeta_e m_H n_e(0)$ and the volume of the star is $\frac{4}{3}\pi R^3$, so in the non-relativistic case where $n_e \sim \epsilon_F^{\frac{3}{2}}$ and $\epsilon_c \sim 1/R^4$,

$$M \propto \epsilon_c^{\frac{3}{2}} R^3 \sim \frac{1}{R^3},  \tag{8.70}$$

and in the ultra-relativistic case where $n_e \sim \epsilon_F^3$,

$$M \propto \epsilon_c^3 R^3 \sim \text{constant}, \tag{8.71}$$

which is independent of $R$, since $\epsilon_c \sim 1/R$ in the ultra-relativistic case.

Consider the thought experiment in which we start with a low-mass, large-radius white dwarf star in which all of the electrons can be treated non-relativistically and then gradually add mass to the star. As the mass increases, the radius will decrease as indicated by Eq. (8.70), and since $\epsilon_F \sim 1/R^4$, the energy of the electrons will increase. If we continue to add mass to the star then at some point, we will find that $\epsilon_F \sim m_e c^2$, the rest mass energy of the electron, at which point it is no longer appropriate to use a non-relativistic treatment of the problem, and we find, as indicated by Eq. (8.71), that the white dwarf has reached a limiting value of the mass as $R \to 0$. This maximum value is known as the Chandrasekhar mass and takes the value

$$M_C = \frac{3.1}{\zeta_e^2 m_H^2} \left( \frac{\hbar c}{G} \right)^{\frac{3}{2}} = \frac{1.45}{(\zeta_e/2)^2} M_\odot, \tag{8.72}$$

where $M_\odot \simeq 2 \times 10^{30}$ kg is the mass of the Sun, and $\zeta_e \simeq 2$ for many likely fully ionized constituents of white dwarf stars. If relativistic effects are not included, one finds that the mass can increase without bound as $R \to 0$ – the behaviour of $M(R)$ is illustrated schematically for both the relativistic and non-relativistic cases in Fig. 8.7.

A white dwarf star with mass larger than $M_C$ is unstable to further gravitational collapse to a neutron star. After such a collapse the degeneracy pressure from neutrons can replace that of electrons to prevent further gravitational collapse. Most observed white dwarf stars have masses between $0.5 M_\odot$ and $M_\odot$. The only known example of a white dwarf star with mass larger than the Chandrasekhar limit was the progenitor star for the supernova SN

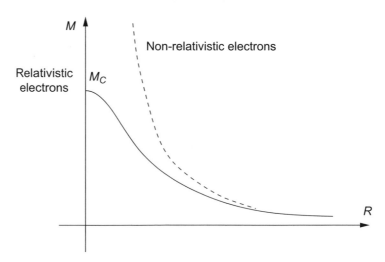

**Fig. 8.7**    Schematic sketch of the mass of a white dwarf as a function of its radius. Note that when relativistic effects are included, there is a maximum mass for the star, the Chandrasekhar mass, $M_C$.

2003fg, which reached a mass of about $2M_\odot$ before exploding. It is possible that it may have been rapidly rotating, which can raise the Chandrasekhar mass.

### 8.3.3 Neutron Stars

As noted above, large progenitor stars can lead to neutron stars for which the pressure opposing gravitational collapse is neutron degeneracy pressure as opposed to electron degeneracy pressure. Neutron stars are even more dense than white dwarfs (masses of order $1$–$2M_\odot$ in a radius of order $R \sim 10$ km) and general relativisitic effects need to be taken into account to describe their properties accurately. Matter in the interior of a neutron star is at temperatures and pressures well beyond those accessible on Earth, so there is considerable theoretical uncertainty as to the structure of their interiors and calculations are considerably more complicated than for white dwarf stars. However, theory suggests that there is also a maximum mass for neutron stars of perhaps at most $3M_\odot$. The largest masses reported for neutron stars are around $2.3M_\odot$, which is consistent with this upper limit.

## 8.4 Summary

The occupation numbers for non-interacting fermions are governed by the Fermi–Dirac distribution

$$f_0(\epsilon) = \frac{1}{e^{\beta(\epsilon-\mu)} + 1}. \tag{8.73}$$

In Fermi systems the chemical potential sets a very important energy scale. At zero temperature the chemical potential is known as the Fermi energy $\epsilon_F$, and all states with energy lower than the Fermi energy are occupied and all states with energy higher than the Fermi energy are unoccupied. The Fermi energy also sets a temperature scale, the Fermi temperature $T_F = \epsilon_F/k_B$. These properties can often greatly simplify the calculation of thermodynamic quantities at low temperatures (i.e. temperatures much less than the Fermi temperature), and give rise to many different phenomena, including the properties of metals and the existence of compact stars such as white dwarfs and neutron stars, which are supported by fermion degeneracy pressure.

## Problems

**8.1** Sodium has an electron density of $n_e = 2.65 \times 10^{28}$ m$^{-3}$. Is a zero-temperature approximation likely to be appropriate for describing the electronic properties of sodium metal at room temperature? Explain your answer.

**8.2** Find the Fermi energy $\epsilon_F$ for a three-dimensional relativistic electron gas in which the energy of the electrons is given by

$$\epsilon^2 = p^2 c^2 + m^2 c^4.$$

Take the non-relativistic limit of your expression for $\epsilon_F$ and comment on the result.

**8.3** Graphene is a single graphitic layer of carbon atoms, which has the remarkable feature that electrons in graphene behave as two-dimensional massless relativistic fermions with a "speed of light" $c_{\text{graphene}} \ll c$. Andre Geim and Konstantin Novoselov were awarded the 2010 Nobel Prize in Physics "for groundbreaking experiments regarding the two-dimensional material graphene".

(a) Calculate the density of states as a function of energy for massless relativistic electrons in two dimensions for a system of size $L$ (note that the degeneracy factor in graphene is $g = 4$).

(b) Calculate the density of states as a function of energy for non-relativistic two-dimensional electrons in two dimensions with spin degeneracy $g_s = 2$ for a system of size $L$.

(c) Use your results from parts (a) and (b) to calculate the Fermi energy $\epsilon_F$ and average energy $U$ at zero temperature when there are $N$ particles for both non-relativistic electrons and relativistic electrons in graphene. Express your answers for $U$ in the form $U = \alpha N \epsilon_F$ and determine $\alpha$.

(d) Find the chemical potential in two dimensions at all temperatures in the non-relativistic case. Show that your expression gives the correct behaviour at low and high temperatures.

**8.4** Find the density of states appropriate for a relativistic Fermi gas in three dimensions, for which

$$\epsilon^2 = p^2 c^2 + m^2 c^4,$$

where $c$ is the speed of light. For the extreme relativistic case ($\epsilon = pc$) obtain expressions for $PV$ and $U$, where $P$ is the pressure, $V$ the volume and $U$ the energy. Without explicitly evaluating these expressions, show that

$$PV = \frac{1}{3} U.$$

Finally show that as $T \to 0$, $U$ approaches the value

$$U = \frac{\pi g_s c p_0^4}{h^3} V,$$

with

$$p_0 = h \left( \frac{3n}{4\pi g_s} \right)^{\frac{1}{3}},$$

where $g_s$ is the spin degeneracy and $n$ is the density.

**8.5** Calculate the density of states as a function of energy for non-relativistic free particles of mass $m$ with spin degeneracy $g_s$ for a system of size $L$:

(a) (i) In one dimension.

(ii) In two dimensions.

(b) Assuming the particles are fermions, use your results from part (a) to calculate the Fermi energy $\epsilon_F$ and average energy $U$ when there are $N$ particles.

(i) In one dimension.

(ii) In two dimensions.

(c) It is possible to confine electrons to a quasi-two-dimensional region in semi-conductor heterostructures made from layers of GaAs and $\mathrm{Al}_{1-x}\mathrm{Ga}_x\mathrm{As}$. If a magnetic field is applied perpendicular to the plane in which the electrons are confined, the energy levels of the electrons are Landau levels, which have the form

$$E_n = \hbar\omega_c \left(n + \frac{1}{2}\right),$$

where $\omega_c = |e|B/m^*$ and $n = 0, 1, 2, \ldots$ labels the Landau levels (which have a very large degeneracy); $|e|$ is the magnitude of the electron charge and $m^* \simeq 0.068 m_e$ is the effective mass of electrons in GaAs, where $m_e$ is the electron mass. If the density of electrons in the two-dimensional layer is $n = 10^{15}$ m$^{-2}$, estimate the magnetic field required so that all electrons are in the lowest Landau level (this is a necessary precondition for observation of the fractional quantum Hall effect).

*Hint:* take your expression for the Fermi energy from part (b)(ii) and compare it to the energy of the $n = 0$ Landau level.

**8.6** (a) Show that in the classical limit we can relate the number $N$ of ideal fermions in three dimensions to a power series in $e^{\beta\mu}$ where the first two terms give

$$N = g_s V n_Q \left[ e^{\beta\mu} - \frac{e^{2\beta\mu}}{2^{\frac{3}{2}}} \right],$$

where $g_s$ is the spin degeneracy and $n_Q$ is the quantum concentration.

*Hint:* you may find it helpful to write

$$\frac{1}{e^{\beta(\epsilon-\mu)} + 1} = \frac{e^{-\beta(\epsilon-\mu)}}{1 + e^{-\beta(\epsilon-\mu)}}.$$

(b) Show that part (a) implies that the chemical potential in the classical limit may be written as $\mu = \mu_0 + \delta\mu$, where

$$\mu_0 = k_B T \ln\left(\frac{n}{g_s n_Q}\right)$$

is the chemical potential for an ideal gas and

$$\delta\mu = k_B T \ln\left[1 + \frac{1}{2^{\frac{3}{2}}}\left(\frac{n}{g_s n_Q}\right)\right].$$

**8.7** A very simple model for an atomic nucleus treats neutrons and protons ("nucleons") as fermions in a three-dimensional simple harmonic oscillator potential. Hence the nucleons have energies

$$\epsilon_{n_x,n_y,n_z} = \hbar\omega \left( n_x + n_y + n_z + \frac{3}{2} \right),$$

with $n_x, n_y, n_z = 0, 1, 2, \ldots$.

(a) Show that the energies above are equivalent to writing the energy as

$$\epsilon_n = \hbar\omega \left( n + \frac{3}{2} \right),$$

with degeneracy

$$g_n = \frac{1}{2}(n + 2)(n + 1).$$

The nucleons have additional degeneracies associated with spin and isospin, so the degeneracy is actually

$$g_n = 2(n + 2)(n + 1).$$

(b) Given the energies and degeneracies in part (a), find the density of states $g(\epsilon)$.
(c) Using your result for part (b), find the Fermi energy $\epsilon_F$ of nucleons in the nucleus, and find the average energy $U$ of the fermions in terms of $\epsilon_F$.

**8.8**  Electrons in a metal can have their spin degeneracy lifted by the application of a magnetic field. For a magnetic field of strength $B$ the electron energy levels are

$$\epsilon_k = \frac{\hbar^2 k^2}{2m} - g\mu_B B\sigma,$$

where $g$ is the Landé $g$ factor, $\mu_B$ the Bohr magneton and $\sigma = \pm 1$ for up (+) and down (−) electron spins, respectively.

The occupation numbers for a given momentum will thus be spin dependent, and hence a gas of electrons develops a magnetic moment

$$\langle M(B,T) \rangle = \sum_{\mathbf{k}} g\mu_B \left( \langle n_{\mathbf{k},\uparrow} \rangle - \langle n_{\mathbf{k},\downarrow} \rangle \right).$$

The magnetic susceptibility is related to the magnetization via

$$\chi(B) = \left. \frac{\partial \langle M \rangle}{\partial B} \right|_{T,N,V},$$

for small magnetic fields

$$\langle M \rangle = \chi(B = 0)B.$$

Obtain the magnetic susceptibility for a non-interacting electron gas in three dimensions and show that in the $T = 0$ limit it takes the form

$$\chi(B = 0) = \frac{3}{2} N \frac{(g\mu_B)^2}{\epsilon_F},$$

and that at high temperatures

$$\chi(B = 0) = N \frac{(g\mu_B)^2}{k_B T}.$$

# Bosons

The most important difference between bosons and fermions, as emphasized in Chapter 7, is that for bosons there is no restriction on the number of particles that can occupy an energy level. We will start by discussing photons as bosons with zero chemical potential, which allows us to derive the thermodynamic properties of blackbody radiation. We then focus on bosons with finite chemical potential, and discuss how Bose–Einstein condensation and superfluidity, examples of macroscopic occupation of a quantum state, can occur at sufficiently low temperatures. Finally we discuss phonons and magnons, which are bosonic excitations that are relevant to the low-temperature properties of solids.

## 9.1 Photons and Blackbody Radiation

In vacuum, the electric field $\mathbf{E}$ satisfies the wave equation

$$\nabla^2 \mathbf{E} - \frac{1}{c^2}\frac{\partial^2 \mathbf{E}}{\partial t^2} = 0, \tag{9.1}$$

where $c$ is the speed of light, with a similar equation for the magnetic field $\mathbf{B}$, which is related to the electric field through Maxwell's equations via

$$\frac{\partial \mathbf{B}}{\partial t} = \nabla \times \mathbf{E}. \tag{9.2}$$

We can use the technique of separation of variables to separate out the spatial and temporal dependence of the solution:

$$\mathbf{E}(\mathbf{x}, t) = \mathbf{X}(\mathbf{x})T(t), \tag{9.3}$$

in which case we get an equation for the spatial part of the solution

$$\nabla^2 \mathbf{X}(\mathbf{x}) = -|\mathbf{k}|^2 \mathbf{X}(x), \tag{9.4}$$

where $|\mathbf{k}|^2$ is a separation constant. The exact form of $\mathbf{X}$ will depend on the boundary conditions – these will determine the allowed values of $\mathbf{k}$, which label the normal modes of the system. They could be travelling plane waves, in which case $\mathbf{k}$ is the wavevector, or modes in a cavity. We also obtain an equation for the temporal part of the solution

$$\ddot{T}(t) = -\omega_k^2 T(t), \tag{9.5}$$

where

$$\omega_k = c|\mathbf{k}| \tag{9.6}$$

is the dispersion relation for the normal modes. There are two modes for each **k**, one for each transverse polarization. In a medium where the propagation velocity is less than the speed of light, there can also be longitudinally polarized modes. Equation (9.5) has exactly the form of the equation of motion of a classical simple harmonic oscillator, so we can view each normal mode as behaving like a simple harmonic oscillator. This harmonic oscillator is not a physical localized oscillator but rather an oscillation of the electric and magnetic fields in the mode.

Thus far our treatment of the electric field has been classical and so it is no surprise that we have obtained the equation for a classical harmonic oscillator. A fully quantum treatment of the electromagnetic field is beyond the scope of our discussion here, but the classical oscillator we have found in Eq. (9.5) generalizes to a quantum SHO in a quantum treatment.[1] We have already studied the quantum SHO in Section 4.4.2, and in Eq. (4.49) we found that the average occupation of a mode with angular frequency $\omega_k$ is given by

$$\langle n \rangle = \frac{1}{e^{\beta \hbar \omega_k} - 1}. \tag{9.7}$$

The quanta of the electromagnetic field are called photons, so Eq. (9.7) counts the number of photons in a mode of the electromagnetic field. It also has exactly the form of the Bose–Einstein distribution for bosons with energy $\epsilon_k = \hbar \omega_k$ and with chemical potential $\mu = 0$. Another way to argue that the chemical potential for photons should be zero is to note that photons can be created or destroyed, and hence the number of photons is not fixed. To an excellent approximation we can treat photons as indistinguishable non-interacting bosons.

In describing photon states mathematically, we will assume that they are plane-wave states in a box of size $L$ with periodic boundary conditions. In the limit that $L$ becomes large, we expect that the exact nature of the boundary conditions will not be important.

### 9.1.1 Blackbody Radiation

Conceptually, when discussing blackbody radiation we consider an ideal surface that emits and absorbs light with 100% efficiency. A model that we can use for this is to imagine a cavity with a small hole (Fig. 9.1). Any radiation that enters the cavity through the hole is very unlikely to be reflected before it has been absorbed by the walls of the cavity. Hence any radiation leaving the hole will come from the thermally excited modes of the cavity and will have a purely thermal spectrum, also known as a blackbody spectrum.

In the preceding section we connected cavity modes with photons, and the results we obtain for photon states will also be applicable to blackbody radiation.

### 9.1.2 Density of States

We can calculate the density of states for photons in a box using the method introduced in Chapter 7. We take into account that photons in vacuum can have two possible transverse polarization states (there can be no longitudinally polarized photon modes in vacuum),

---

[1] Most graduate-level quantum mechanics textbooks give a discussion of the quantization of the electromagnetic field.

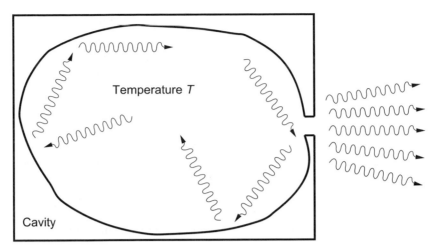

Fig. 9.1 A blackbody cavity with temperature $T$. Photons are emitted from a small hole in the cavity.

i.e. the degeneracy factor $g = 2$, to find the number of states in a thin shell of thickness $dk$ in $k$-space as

$$dN = 2\frac{4\pi k^2 dk}{\left(\frac{2\pi}{L}\right)^3}$$

$$= \frac{Vk^2 dk}{\pi^2}$$

$$= \frac{V\omega^2 d\omega}{\pi^2 c^3}, \tag{9.8}$$

where we used the dispersion Eq. (9.6) to go from the second to the third line above. Hence the density of plane waves per unit frequency is[2]

$$g(\omega) = \frac{dN}{d\omega} = \frac{V\omega^2}{\pi^2 c^3}. \tag{9.9}$$

### 9.1.3 Number Density

The number of photons with energy between $\omega$ and $\omega + d\omega$, $N(\omega)d\omega$, is given by the product of the density of states and the occupation numbers from the Bose–Einstein distribution:

$$N(\omega)d\omega = \frac{g(\omega)}{e^{\beta\hbar\omega} - 1}d\omega. \tag{9.10}$$

This implies a number density of photons at temperature $T$ of

$$\langle n \rangle = \frac{N}{V}$$

$$= \frac{1}{V}\int_0^\infty d\omega\, N(\omega)$$

[2] We will sometimes refer to frequency when discussing angular frequency $\omega = 2\pi f$, where $f$ is the frequency. We will be consistent in notation that $\omega$ is always an angular frequency.

$$= \frac{1}{\pi^2 c^3} \int_0^\infty d\omega \frac{\omega^2}{e^{\beta\hbar\omega} - 1}$$

$$= \frac{1}{\pi^2} \left(\frac{k_B T}{\hbar c}\right)^3 \int_0^\infty dx \frac{x^2}{e^x - 1}$$

$$= \frac{1}{\pi^2} \left(\frac{k_B T}{\hbar c}\right)^3 \Gamma(3)\zeta(3)$$

$$= \frac{2.404}{\pi^2} \left(\frac{k_B T}{\hbar c}\right)^3, \tag{9.11}$$

where we noted that $\Gamma(3) = 2!$ and $\zeta(3) = 1.202$, where $\zeta(s)$ is the Riemann zeta function defined as

$$\zeta(s) = \sum_{n=1}^\infty \frac{1}{n^s}. \tag{9.12}$$

The integral over $x$ that arises in the derivation of Eq. (9.11) is a special case of a more general class of integral that appears in many calculations of the properties of bosons:

$$\int_0^\infty dx \frac{x^m}{e^x - 1} = \int_0^\infty dx \frac{x^m e^{-x}}{1 - e^{-x}}$$

$$= \sum_{n=0}^\infty \int_0^\infty dx \, x^m e^{-(n+1)x}$$

$$= \sum_{n=0}^\infty \frac{1}{(n+1)^{m+1}} \int_0^\infty dy \, y^m e^{-y}$$

$$= \Gamma(m+1) \sum_{n=1}^\infty \frac{1}{n^{m+1}}$$

$$= \Gamma(m+1)\zeta(m+1), \tag{9.13}$$

where we made a change of variable from $x$ to $y = (n+1)x$ (note that our derivation does not assume that $m$ is an integer). In the calculation of $\langle n \rangle$ in Eq. (9.11) we considered $m = 2$.

### 9.1.4 Energy Density

The energy per unit volume $u(\omega)$ in the frequency interval $[\omega, \omega+d\omega]$ is the spectral energy density and it satisfies

$$V u(\omega) d\omega = \hbar\omega N(\omega) d\omega$$

$$= \frac{\hbar\omega g(\omega)}{e^{\beta\hbar\omega} - 1} d\omega$$

$$= \frac{V\hbar}{\pi^2 c^3} \frac{\omega^3 d\omega}{e^{\beta\hbar\omega} - 1}. \tag{9.14}$$

Physically this describes the frequency dependence of the thermal emission spectrum of an object such as a star, a piece of red-hot metal or a human being. If we just focus on $u(\omega)$, this is Planck's expression for the energy density in thermal (blackbody) radiation

$$u(\omega) = \frac{\hbar}{\pi^2 c^3} \frac{\omega^3}{e^{\beta\hbar\omega} - 1}. \tag{9.15}$$

In 1900 Max Planck obtained the spectrum Eq. (9.15) to match experimental observations by treating a cavity as a blackbody and making the assumption that the cavity modes can only take discrete values $n\epsilon$ for integers $n$ (in a more modern picture this is equivalent to saying there are $n$ photons in the mode with energy $\epsilon$). He took the energy of the mode to be $\epsilon = hf$, with $f$ the frequency and $h = 6.626 \times 10^{-34}$ J s a constant of proportionality now known as Planck's constant. We will generally work with the angular frequency $\omega = 2\pi f$, in which case $\epsilon = \hbar\omega$.

The maximum of the spectral energy density in Eq. (9.15) occurs when

$$\frac{du(\omega)}{d\omega} = 0, \tag{9.16}$$

and if we write $x = \beta\hbar\omega$, then we need to find the maximum of

$$\frac{x^3}{e^x - 1},$$

which satisfies

$$3 - 3e^{-x} = x, \tag{9.17}$$

which can be solved numerically to obtain $\omega_{max}$, the frequency at which $u(\omega)$ is maximum:

$$\frac{\hbar\omega_{max}}{k_B T} = 2.82, \tag{9.18}$$

which is **Wien's displacement law**. The significance of this result is that the temperature of a blackbody can be extracted from the maximum in the spectral density:

$$T = \frac{\hbar\omega_{max}}{2.82 k_B}. \tag{9.19}$$

To get the total energy density per unit volume we can integrate $u(\omega)$ over all frequencies:

$$u = \int_0^\infty d\omega\, u(\omega)$$

$$= \frac{\hbar}{\pi^2 c^3} \int_0^\infty d\omega \frac{\omega^3}{e^{\beta\hbar\omega} - 1}$$

$$= \frac{(k_B T)^4}{\pi^2 c^3 \hbar^3} \int_0^\infty dx \frac{x^3}{e^x - 1}$$

$$= \Gamma(4)\zeta(4)\frac{(k_BT)^4}{\pi^2 c^3 \hbar^3}$$

$$= \frac{\pi^2}{15}\frac{(k_BT)^4}{(\hbar c)^3},\tag{9.20}$$

where we noted that the integral is of the form of Eq. (9.13) with $m = 3$, and $\Gamma(4) = 3! = 6$ and[3]

$$\zeta(4) = \sum_{n=1}^{\infty}\frac{1}{n^4} = \frac{\pi^4}{90}.\tag{9.21}$$

The approach we took to the integrals in Eqs (9.11) and (9.20) was to make them dimensionless, so that we were able to capture the temperature dependence of the energy density up to a numerical factor. In these cases the numerical factors can be expressed in terms of known special functions, but more commonly these may need to be evaluated numerically.

If we consider the spectral energy density of Eq. (9.15) at low frequencies, we can expand

$$e^{\beta\hbar\omega} - 1 \simeq \frac{\hbar\omega}{k_BT},\tag{9.22}$$

which leads to the **Rayleigh–Jeans formula** for the energy density

$$Vu_{RJ}(\omega)d\omega = V\left(\frac{k_BT}{\pi^2 c^3}\right)\omega^2 d\omega = k_BTg(\omega)d\omega.\tag{9.23}$$

An alternative approach to obtaining this result is to use classical equipartition, which gives $k_BT$ per mode, and multiply by the number of modes in the frequency window $d\omega$, which is $g(\omega)d\omega$. The Rayleigh–Jeans result becomes inaccurate at frequencies where classical equipartition does not give a good estimate for the average energy of the mode.

The Rayleigh–Jeans formula implies that at large frequencies, the energy density increases without bound, which is clearly unphysical. This unphysical behaviour, that failed to match the experimentally observed blackbody spectrum, was a big challenge to nineteenth-century physics. Planck's discovery that quantizing the energy leads to a physically sensible answer that matches experiment and gives an energy density that is finite at high frequencies was a radical idea at the time and an important step towards the development of quantum theory. Planck was awarded the 1918 Nobel Prize in Physics for this work. The difference between the Planck and the Rayleigh–Jeans energy densities is illustrated in Fig. 9.2.

We can note that with knowledge of the number density and the energy density, we can calculate the mean energy per photon, which is

$$\frac{u}{n_\gamma} = \frac{\frac{\pi^2(k_BT)^4}{15(\hbar c)^3}}{\frac{2.404(k_BT)^3}{\pi^2(\hbar c)^3}} = \frac{\pi^4}{15\times 2.404}k_BT = 2.701\,k_BT,\tag{9.24}$$

[3] Equation (9.21) can be derived by considering the Fourier series for $x^4$ on the interval $[-\pi, \pi]$.

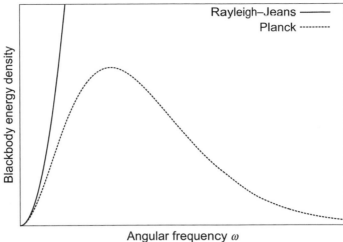

**Fig. 9.2**    Comparison of Rayleigh–Jeans and Planck blackbody energy densities.

so we can see that the temperature is the sole parameter that sets the energy scale for the mean photon energy.

### 9.1.5 Example: Cosmic Microwave Background Radiation

The cosmic microwave background (CMB) radiation is a remnant from the early history of the universe, dating back to when the universe was about 400 000 years old. In 1964, Arno Penzias and Robert Wilson, at Bell Labs in New Jersey, working with a 6-m horn antenna, observed what they believed to be isotropic noise from all parts of the sky. After trying to eliminate all potential sources of noise in their antenna, including pigeons nesting inside it, they became aware of predictions that there should be uniform cosmic microwave background radiation with a blackbody spectrum. This spectrum is illustrated in Fig. 9.3. Penzias and Wilson were awarded the 1978 Nobel Prize in Physics for their discovery.

There have been many subsequent measurements of CMB radiation, as illustrated in Fig. 9.4, and it is now known to be very close to isotropic and to have a temperature of $\sim 2.7$ K. These more detailed measurements, including anisotropies in the CMB, are used to place constraints on cosmological models.

We can use Eq. (9.11) to calculate the photon density for this radiation to be

$$n_\gamma \sim \frac{2.404}{\pi^2} \left( \frac{2\pi \times 1.3807 \times 10^{-23} \times 2.7}{6.626 \times 10^{-34} \times 3 \times 10^8} \right)^3 = 4 \times 10^8 \, \text{m}^{-3}.$$

We can also use Eq. (9.24) to calculate the average energy per microwave photon:

$$\frac{u}{n_\gamma} = 2.701 \times 2.7 k_B \simeq 10^{-22} \, \text{J}.$$

This compares to an energy of roughly $4 \times 10^{-19}$ J per photon for sunlight.

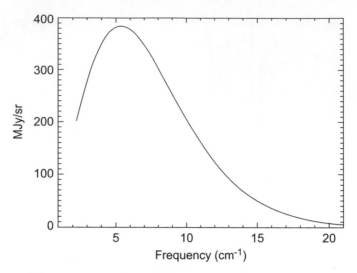

Fig. 9.3 Blackbody spectrum of the cosmic microwave background radiation as measured by the COBE satellite (Fixsen *et al.*, 1996). The difference between the Planck energy density, Eq. (9.15), and the experimental data is too small to discern.

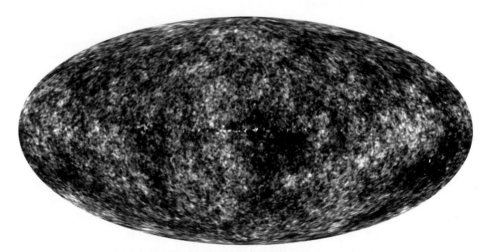

Fig. 9.4 All sky map of the cosmic microwave background radiation as determined by the Wilkinson Microwave Anisotropy Probe (WMAP). Credit: NASA/WMAP Science Team.

### 9.1.6 Radiation Pressure

We can also ask what pressure a thermal photon gas exerts on the walls of its box. Recall that

$$P = \left.\frac{\partial \Phi}{\partial V}\right|_{\mu,T}, \tag{9.25}$$

and remembering that $\mu = 0$ for photons, we can see that

$$P = -k_B T \frac{\partial}{\partial V} \sum_s \ln \left(1 - e^{-\beta \epsilon_s}\right)$$

$$= -k_B T \sum_s \frac{\partial \epsilon_s}{\partial V} \frac{\beta e^{-\beta \epsilon_s}}{1 - e^{-\beta \epsilon_s}}$$

$$= -\sum_s \frac{\partial \epsilon_s}{\partial V} \frac{e^{-\beta \epsilon_s}}{1 - e^{-\beta \epsilon_s}}. \tag{9.26}$$

Now, we can calculate $\partial \epsilon_s / \partial V$ by recalling that the normal modes in a box have wavenumbers $k_s = 2\pi n_s / L$ indexed by $n_s$, so we obtain

$$\epsilon_s = \hbar \omega_s = \hbar c k_s = \hbar c \frac{2 n_s \pi}{L} = \frac{2 \hbar c n_s \pi}{V^{\frac{1}{3}}}, \tag{9.27}$$

and hence

$$\frac{\partial \epsilon_s}{\partial V} = -\frac{\epsilon_s}{3V}, \tag{9.28}$$

so we have

$$P = \frac{1}{3V} \sum_s \frac{\epsilon_s}{e^{\beta \epsilon_s} - 1} = \frac{U}{3V} = \frac{u}{3}. \tag{9.29}$$

Thus the pressure is one-third of the energy density, and since the energy density scales as $T^4$, so does the radiation pressure.

### 9.1.7 Stefan–Boltzmann Law

We introduced the idea of a blackbody in Section 9.1.1 and we now use this picture to determine the radiation rate from a blackbody. We consider the emission from a cavity containing a photon gas in equilibrium as shown in Fig. 9.1. We saw in Chapter 5 for the ideal gas (Section 5.3.2) that effusion was proportional to the flux of particles incident on a hole, and to calculate the power per unit area emitted from a cavity, we will take an analogous approach. Using similar arguments to those we used for effusion of an ideal gas, the number of photons per unit area per unit time with momentum in the range $[\mathbf{p}, \mathbf{p} + d\mathbf{p}]$ exiting through the hole shown in Fig. 9.1 is

$$c \cos \theta \, n_\gamma(\mathbf{p}) p^2 dp \sin \theta d\theta d\phi,$$

where we restrict to photons with $0 < \theta < \frac{\pi}{2}$ (since only photons moving towards the hole can escape through it). To obtain the power radiated per unit area $J$ we need to multiply the photon flux by photon energy and integrate over all appropriate momentum values:

$$J = \int_+ d^3\mathbf{p}(pc)c \cos \theta n_\gamma(\mathbf{p})$$

$$= c \int_0^{2\pi} d\phi \int_0^{\frac{\pi}{2}} d\theta \, \sin \theta \cos \theta \left[ \frac{1}{4\pi} \int_0^\infty dE \, E \, n_\gamma(E) \right]$$

$$\begin{aligned}
&= 2\pi c \left[ -\frac{1}{2} \cos(2\theta) \right]_0^{\frac{\pi}{2}} \frac{u}{4\pi} \\
&= \frac{c}{4} u \\
&= \frac{c}{4} \frac{\pi^2}{15} \frac{(k_B T)^4}{(\hbar c)^3} \\
&= \sigma_{SB} T^4.
\end{aligned} \tag{9.30}$$

This is the **Stefan–Boltzmann** law, and

$$\sigma_{SB} = \frac{\pi^2 k_B^4}{60 \hbar^3 c^2} = 5.67 \times 10^{-8} \, \text{W} \, \text{m}^{-2} \, \text{K}^{-4} \tag{9.31}$$

is the *Stefan–Boltzmann constant*. In our derivation of Eq. (9.30) we noted that $E = pc$ for photons and hence made the observation that

$$\int d^3 \mathbf{p} (pc) n_\gamma(\mathbf{p}) = 4\pi \int_0^\infty p^2 \, dp \, (pc) \, n_\gamma(\mathbf{p}) = \int_0^\infty dE \, E \, n_\gamma(E). \tag{9.32}$$

The calculation of the Stefan–Boltzmann law assumed a perfect blackbody, which absorbs all incident radiation and emits at the same rate (if it didn't, its energy, and hence its temperature $T$, would change and it would not be in equilibrium). However, the perfect blackbody is an idealization not actually realized physically – all physical systems are imperfect blackbodies. Imperfect blackbodies must still emit heat at the same rate they absorb it in order that their temperature stays constant. We can modify the Stefan–Boltzmann law for an imperfect blackbody to

$$J = \varepsilon \sigma_{SB} T^4, \tag{9.33}$$

where $\varepsilon \leq 1$ is the emissivity and measures the level to which a blackbody is imperfect. For a perfect blackbody, $\varepsilon = 1$. In general, $\varepsilon$ may be a function of the frequency of radiation from the blackbody.

## 9.2 Bose–Einstein Condensation

For bosons that cannot be created or destroyed, and hence have a non-zero chemical potential, we can determine the total number of particles by integrating over the Bose–Einstein distribution:

$$N = \int_0^\infty d\epsilon \, g(\epsilon) \frac{1}{e^{\beta(\epsilon - \mu)} - 1}. \tag{9.34}$$

From our discussion in Chapter 7, we know that for the Bose–Einstein distribution the chemical potential $\mu$ must always be less than the energy of the lowest-energy orbital, $\epsilon_0$, which we take to be $\epsilon_0 = 0$, in order for $\langle n(\epsilon) \rangle$ to be finite. We obtained Eq. (9.34) using the grand canonical ensemble, to find $N$ as a function of $\mu$. If we are considering a system with a fixed number of bosons (as is usually the case in experiments), then if we change thermodynamic variables such as temperature, we need to consider how the chemical potential changes as a function of those variables. As the temperature is lowered, $\mu$ needs to change so that Eq. (9.34) is satisfied. The only place that temperature and $\mu$ enter the integral is in the Bose–Einstein distribution, and we saw the dependence of this distribution on $\beta(\epsilon - \mu)$ in Fig. 7.3. For a fixed number of particles, $N$, this means that any changes in temperature (i.e. $\beta$) must be accommodated by changes in $\epsilon - \mu$. As temperature is decreased, $\beta$ increases, which means that in order to keep $N$ fixed, $\epsilon - \mu$ must decrease. Hence, as the temperature is lowered, we expect $\mu$ to increase and to approach $\epsilon_0$ from below. At some sufficiently low temperature, which we shall call $T_0$, the chemical potential $\mu \sim 0$. We can relate $T_0$ to the number of particles $N$ (the integral converges in three dimensions) via

$$
\begin{aligned}
N &= \int_0^\infty d\epsilon \frac{g(\epsilon)}{e^{\beta_0 \epsilon} - 1} \\
&= \frac{g_s V}{4\pi^2} \left( \frac{2m}{\hbar^2} \right)^{\frac{3}{2}} \int_0^\infty d\epsilon \frac{\sqrt{\epsilon}}{e^{\beta_0 \epsilon} - 1} \\
&= \frac{g_s V}{4\pi^2} \left( \frac{2m}{\hbar^2} \right)^{\frac{3}{2}} (k_B T_0)^{\frac{3}{2}} \int_0^\infty dx \frac{\sqrt{x}}{e^x - 1} \\
&= \frac{g_s V}{4\pi^2} \left( \frac{2m k_B T_0}{\hbar^2} \right)^{\frac{3}{2}} \Gamma\left(\frac{3}{2}\right) \zeta\left(\frac{3}{2}\right) \\
&= g_s V \left( \frac{m k_B T_0}{2\pi\hbar^2} \right)^{\frac{3}{2}} \zeta\left(\frac{3}{2}\right) \\
&= \zeta\left(\frac{3}{2}\right) g_s n_Q^0 V,
\end{aligned}
\tag{9.35}
$$

where $\beta_0 = 1/k_B T_0$ and $n_Q^0$ is the quantum concentration $n_Q$ evaluated at the temperature $T_0$. The integral over $x$ is of the form of Eq. (9.13), and takes the value

$$
\int_0^\infty dx \frac{\sqrt{x}}{e^x - 1} = \Gamma\left(\frac{3}{2}\right) \zeta\left(\frac{3}{2}\right).
\tag{9.36}
$$

We recall that $\Gamma\left(\frac{3}{2}\right) = \sqrt{\pi}/2$ and note that $\zeta\left(\frac{3}{2}\right) \simeq 2.612$.

The observation that $\mu < \epsilon_0$, coupled with Eq. (9.35), raises the question of what happens in a system with fixed $N$ when we go to temperatures $T < T_0$. The answer lies in how we have been performing the calculation of the number of particles in the system. In the continuum approach, the density of states at zero energy is $g(0) = 0$. This means that at low enough temperatures when there is a sizeable population of bosons in the lowest-energy orbital, the continuum approach can become qualitatively wrong. If we repeated the calculation of $N$ using the continuum approach for some $T < T_0$ where we also have $\mu \simeq 0$, we would get a different value for $N$, which clearly cannot be correct as it is not compatible with a fixed number of bosons. The bosons we have not accounted for must be accommodated by the lowest-energy orbital. At temperatures $T < T_0$ we should thus write

$$N = \langle N_0 \rangle + g_s V \left( \frac{mk_B T}{2\pi\hbar^2} \right)^{\frac{3}{2}} \zeta\left(\frac{3}{2}\right), \tag{9.37}$$

where $\langle N_0 \rangle$ is the average occupation of the lowest-energy orbital (which will be temperature dependent). We can rearrange Eq. (9.37) using our expression for $N$ in terms of $T_0$ as

$$\frac{N - \langle N_0 \rangle}{N} = \left( \frac{T}{T_0} \right)^{\frac{3}{2}}, \tag{9.38}$$

and hence the fraction of the bosons that are in the lowest-energy orbital, the *condensate fraction* (see Fig. 9.5), is

$$\frac{\langle N_0 \rangle}{N} = 1 - \left( \frac{T}{T_0} \right)^{\frac{3}{2}}. \tag{9.39}$$

This is the phenomenon of Bose–Einstein condensation in which the lowest-energy orbital has macroscopic occupation below a temperature $T_0$, and as the temperature is lowered, an increasing proportion of the bosons are in the lowest-energy orbital, with all bosons in the lowest-energy orbital when $T = 0$. We can obtain $T_0$ by inverting Eq. (9.35) to get the critical temperature for Bose–Einstein condensation:

$$T_0 = \frac{2\pi\hbar^2}{mk_B} \left( \frac{n}{g_s \zeta\left(\frac{3}{2}\right)} \right)^{\frac{2}{3}}. \tag{9.40}$$

If $g_s = 1$, then we can write this as

$$T_0 = 3.31 \frac{\hbar^2}{mk_B} n^{\frac{2}{3}}. \tag{9.41}$$

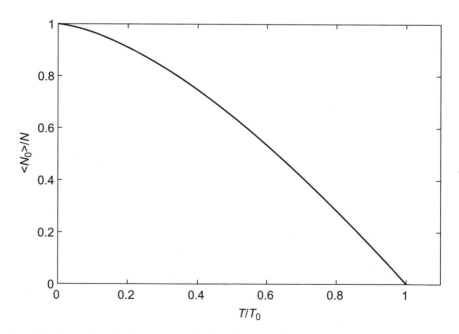

Fig. 9.5   Condensate fraction as a function of temperature in a Bose–Einstein condensate.

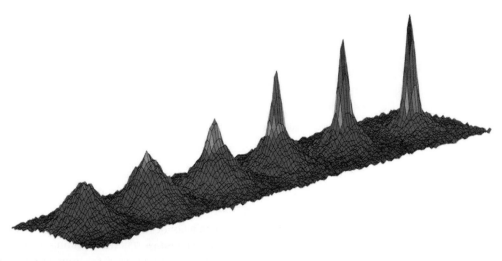

Density profile of atoms in a confining potential as the temperature is lowered. At temperatures $T > T_0$ when all of the atoms are in the normal fraction there is a Gaussian density profile, and at temperatures $T < T_0$ when a Bose–Einstein condensate has formed there is a sharp peak associated with the condensate fraction and a smaller normal background. Reproduced courtesy of NASA/JPL-Caltech.

Alternatively, we can view this as a critical density

$$n_0 = g_s \zeta \left( \frac{3}{2} \right) n_Q = 2.612 \, g_s n_Q, \tag{9.42}$$

and Bose–Einstein condensation is possible for densities $n > n_0$. At temperatures below $T_0$, we can view the system as being composed of two components: a condensate fraction $N_0/N$ and a normal fraction $1 - (N_0/N)$. Bose–Einstein condensation of atoms in a trap was achieved in 1995, and the 2001 Nobel Prize was awarded to Wolfgang Ketterle, Carl Wieman and Eric Cornell for this achievement.

It is important to realize that Bose–Einstein condensation is a phase transition that arises because of particle statistics – it does not require interactions, and occurs due to a condensation in phase space (to the $\mathbf{k} = 0$ state). It is **not** a condensation in position space; however, Bose–Einstein condensation of a gas in a confining potential can lead to a change in the spatial distribution of atoms, as illustrated in Fig. 9.6.

That Bose–Einstein condensation is a statistical phenomenon can be seen by taking the very naive view that we might be able to estimate the temperature at which the phase transition occurs by comparing the energies of the two lowest-energy states. If we consider bosons in a three-dimensional box of size $L$, the energy levels are given by

$$E_{n_x,n_y,n_z} = \frac{\hbar^2 \pi^2}{2mL^2} (n_x^2 + n_y^2 + n_z^2), \tag{9.43}$$

and the difference in energy between the two lowest states is

$$\Delta E = E_{2,1,1} - E_{1,1,1} = \frac{3\hbar^2 \pi^2}{2mL^2}, \tag{9.44}$$

and for $^{87}$Rb, which has $m = 1.45 \times 10^{-25}$ kg, in a box of size $L = 1$ cm (similar in size to the trap used experimentally), this implies $\Delta E \simeq 10^{-38}$ J, which corresponds to a temperature of $T = \Delta E/k_B \simeq 8 \times 10^{-16}$ K. This temperature scale is well below the 170 nK temperature at which Bose–Einstein condensation was first observed (Anderson *et al.*, 1995), which we can see from our earlier considerations depends on the density, whereas this naive estimate in Eq. (9.44) does not.

### 9.2.1 Superfluidity

At a temperature of 2.176 K, liquid $^4$He has a transition to a state which lacks viscosity: a superfluid. We can calculate the Bose–Einstein condensation temperature using Eq. (9.41) with parameters appropriate for $^4$He to be 3.13 K, which suggests that Bose–Einstein condensation may be related to superfluidity. Our calculation of the Bose–Einstein condensation temperature in Eq. (9.41) was for non-interacting bosons, whereas the interactions between helium atoms cannot be neglected for a quantitative description of the transition to a superfluid.

We will not attempt to include the interactions between the particles, but we can give a simple discussion of some of the properties of a superfluid following the ideas first put forward by Landau. To get an understanding of the lack of viscosity in a superfluid, consider a massive object moving through a fluid with some velocity. In a regular fluid the object collides with particles in the fluid, which leads to fluctuations in the density that propagate as sound waves. Energy and momentum are transferred from the object to the fluid, so that the object effectively feels a viscous force that retards its motion. If we consider the same experiment for a superfluid, which is at close to zero temperature, then the quanta of sound waves are phonons (elementary excitations of the liquid) with momentum $\hbar\mathbf{k}$ and energy $\epsilon_k$ and again can lead to a viscous force. However, due to the nature of the excitations in a superfluid, the viscous force vanishes unless the object is moving faster than a critical speed. We will justify this phenomenology below.

For liquid $^4$He at rest at zero temperature, there are no elementary excitations. When we introduce an object with mass $M_0$ and velocity $\mathbf{V}_0$, we can expect that the motion of the mass transfers energy to the helium. This energy must put the helium into an excited state, in which there is some density of excitations with energy $\epsilon_k$ and momentum $\hbar\mathbf{k}$. These excitations must satisfy conservation of energy:

$$\frac{1}{2}M_0V_0^2 = \frac{1}{2}M_0V_1^2 + \epsilon_k \tag{9.45}$$

and momentum:

$$M_0\mathbf{V}_0 = M_0\mathbf{V}_1 + \hbar\mathbf{k}, \tag{9.46}$$

where $\mathbf{V}_1$ is the velocity of the mass after an elementary excitation has been created. The key for superfluidity is that these two conditions cannot always be satisfied at the same time for the form of $\epsilon_k$ found in liquid $^4$He. To see this, square the momentum equation, Eq. (9.46), to get

$$\frac{1}{2}M_0V_1^2 = \frac{1}{2M_0}(M_0\mathbf{V}_0 - \hbar\mathbf{k})^2 = \frac{1}{2}M_0V_0^2 - \hbar\mathbf{V}_0 \cdot \mathbf{k} + \frac{\hbar^2k^2}{2M_0}, \tag{9.47}$$

which we can use to write the energy equation, Eq. (9.45), as

$$\epsilon_k = \hbar \mathbf{V}_0 \cdot \mathbf{k} - \frac{\hbar^2 k^2}{2M_0}. \tag{9.48}$$

The smallest value of the velocity $\mathbf{V}_0$ for which this can be satisfied will occur when $\mathbf{V}_0$ is parallel to $\mathbf{k}$, and this critical speed is

$$V_{\text{crit}} = \min \left[ \frac{\epsilon_k + \frac{\hbar^2 k^2}{2M_0}}{\hbar k} \right], \tag{9.49}$$

and if we consider the limit in which $M_0$ is very large, we get

$$V_{\text{crit}} = \min \left[ \frac{\epsilon_k}{\hbar k} \right]. \tag{9.50}$$

If the speed of a body in the superfluid is less than $V_{\text{crit}}$, it cannot create excitations in the liquid, and hence will experience zero viscosity. In the frame of the superfluid in a container we can consider the walls of a container as moving with respect to the liquid – excitations can be created by the interactions between the liquid and the walls of the container. In the laboratory frame we would consider the superfluid moving with respect to the container but if there are no excitations, then there is no viscosity. The result above means that a superfluid will flow without viscosity provided its speed $v$ in the laboratory frame is less than $V_{\text{crit}}$.

If $\epsilon_k = \frac{\hbar^2 k^2}{2m}$, then $V_{\text{crit}} = 0$, corresponding to $k = 0$. However, in $^4$He the spectrum of excitations has a different form. For small $k$, $\epsilon_k = \hbar v_s k$, where $v_s$ is the sound speed and for larger values of $k$ there is a local minimum known as the roton minimum located at a wavevector $k_0$. It is found experimentally that $V_{\text{crit}}$ is actually less than $v_s$, as illustrated schematically in Fig. 9.7. The minimum in the dispersion for $^4$He is at a wavevector $k_0$ and corresponds to an energy $\Delta$, leading to

$$V_{\text{crit}} = \frac{\Delta}{\hbar k_0} \simeq 50 \, \text{m s}^{-1}. \tag{9.51}$$

The analysis we presented above only applies to a superfluid. As is the case in a Bose–Einstein condensate, at non-zero temperatures, some fraction of the system will be in the normal phase (the normal fraction) in addition to the condensate fraction. The normal fraction, not being condensed, will provide viscosity at any temperature.

## Superfluid $^3$He

Helium-4 is not the only superfluid. In 1972, Douglas Osheroff, Robert Richardson and David Lee (Osheroff *et al.*, 1972) discovered that the helium isotope $^3$He also shows superfluid phases at much lower temperatures than $^4$He, on the order of milli-Kelvins (as compared to $\sim 2$ K in $^4$He). They shared the 1996 Nobel Prize in Physics for this discovery. The difference in temperature scales between $^3$He and $^4$He arises because $^3$He is a fermionic atom and the bosons that condense are not simply atoms, but pairs of atoms (that form at low temperatures). A similar phenomenon of pairing of fermions into bosons that can condense arises when electrons form Cooper pairs leading to superconductivity in metals.

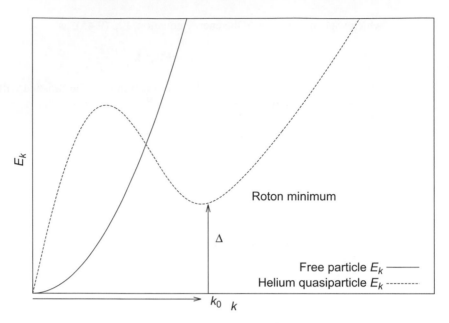

**Fig. 9.7**  Schematic comparison of elementary excitation spectrum of superfluid $^4$He with a free particle dispersion.

## 9.3 Low-Temperature Properties of a Bose Gas

We saw in Chapter 8 that there is a characteristic temperature below which quantum degeneracy becomes important in fermion systems, the Fermi temperature. The analogous temperature in bosonic systems is the Bose–Einstein condensation temperature $T_0$. For temperatures less than $T_0$, the picture of a condensate fraction and a normal fraction allows us to differentiate whether the excited states or ground state contribute to thermodynamic quantities.

### 9.3.1 Chemical Potential

When we have significant occupation of the ground state, making the approximation that $\mu = 0$ is not good enough, since it implies $\langle N_0 \rangle$ is infinite. What we can do to remedy this is to note that

$$\langle N_0 \rangle = \frac{1}{e^{-\beta\mu} - 1}, \tag{9.52}$$

and we know that $\beta\mu \to 0$ as $T \to 0$, so expand the exponential

$$e^{-\beta\mu} \simeq 1 - \beta\mu, \tag{9.53}$$

and then

$$\langle N_0 \rangle \simeq -\frac{k_B T}{\mu}, \tag{9.54}$$

or

$$\mu = -\frac{k_B T}{\langle N_0 \rangle}, \tag{9.55}$$

and in the limit $T \to 0$, $N_0 \to N$. This can clearly be very close to zero when $N$ is large and $k_B T$ is small. We can use Eq. (9.55) to test the claim that there is macroscopic occupation of the lowest-energy state. Consider the first excited particle in a box state, which will have energy

$$\epsilon_1 = \frac{\hbar^2}{2m} \left( \frac{2\pi}{L} \right)^2 \propto \frac{1}{V^{\frac{2}{3}}} = \frac{n^{\frac{2}{3}}}{N^{\frac{2}{3}}}. \tag{9.56}$$

Now, if there is macroscopic occupation of the lowest-energy orbital, then we can write $\langle N_0 \rangle = \alpha N$, for some constant $\alpha$, so $\beta\mu = -1/\alpha N$ and from Eq. (9.56) we can see that $\beta\epsilon_1 = b/N^{\frac{2}{3}}$ for some constant $b$. Thus, when we use the Bose–Einstein distribution to calculate the occupation of the first excited state we get

$$\begin{aligned}
\langle N_1 \rangle &= \frac{1}{e^{\beta(\epsilon_1 - \mu)} - 1} \\
&= \frac{1}{e^{bN^{-\frac{2}{3}} + \frac{1}{\alpha}N^{-1}} - 1} \\
&\simeq \frac{1}{1 + bN^{-\frac{2}{3}} + \frac{1}{\alpha}N^{-1} - 1} \\
&= \frac{N^{\frac{2}{3}}}{b + \frac{1}{\alpha}N^{-\frac{1}{3}}} \\
&\simeq \frac{N^{\frac{2}{3}}}{b}, \tag{9.57}
\end{aligned}$$

where we simplified the expression in the $N \to \infty$ limit. This allows us to compare the relative occupation of the first excited state to the ground state:

$$\frac{\langle N_1 \rangle}{\langle N_0 \rangle} \sim \frac{1}{N^{\frac{1}{3}}} \to 0, \tag{9.58}$$

as $N \to \infty$, which confirms that in the thermodynamic limit there is macroscopic occupation of only the lowest-energy state. The chemical potential in a Bose gas as a function of temperature, and its comparison to a Fermi gas, is shown in Fig. 9.8.

## 9.3.2  Internal Energy and Heat Capacity

We will not calculate the pressure for an ideal gas of bosons explicitly here, but by using similar reasoning to that applied for fermions in Section 8.2.2, we can determine that

$$P = \frac{2}{3}\frac{U}{V}, \tag{9.59}$$

for non-relativistic bosons in three dimensions at all temperatures. The behaviour we discuss below for the internal energy $U$ implicitly tells us the behaviour of $P$.

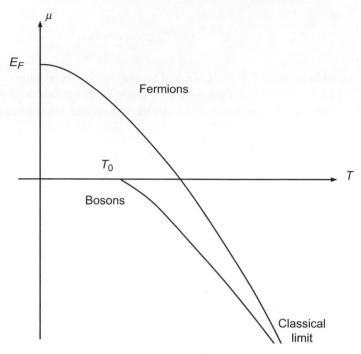

Fig. 9.8 Comparison of the chemical potential of fermion and boson systems as a function of temperature.

We saw in Eq. (9.37) that at temperatures below $T_0$ the number of bosons in excited states is

$$N_{\text{ex}}(T) = N \left(\frac{T}{T_0}\right)^{\frac{3}{2}},$$ (9.60)

and only the excited-state bosons will contribute to the energy, which is (writing the Bose–Einstein distribution as $f_B(\epsilon)$ and taking $\mu = 0$)

$$
\begin{aligned}
U &= \int_0^\infty d\epsilon\, g(\epsilon)\epsilon f_B(\epsilon) \\
&= \frac{g_s V}{4\pi^2} \left(\frac{2m}{\hbar^2}\right)^{\frac{3}{2}} \int_0^\infty d\epsilon \frac{\epsilon^{\frac{3}{2}}}{e^{\beta\epsilon} - 1} \\
&= k_B T N_{\text{ex}}(T) \frac{\Gamma\left(\frac{5}{2}\right)\zeta\left(\frac{5}{2}\right)}{\Gamma\left(\frac{3}{2}\right)\zeta\left(\frac{3}{2}\right)} \\
&= 0.770 N_{\text{ex}}(T) k_B T \\
&= 0.77 N k_B T \left(\frac{T}{T_0}\right)^{\frac{3}{2}}.
\end{aligned}
$$ (9.61)

This leads naturally to an expression for the heat capacity at constant volume for $T < T_0$:

$$C_V = 1.93 N k_B \left(\frac{T}{T_0}\right)^{\frac{3}{2}}.$$ (9.62)

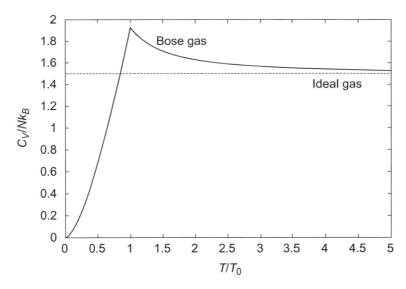

Fig. 9.9  Heat capacity of a Bose gas and a classical ideal gas.

We can obtain an expression for the heat capacity at $T > T_0$ by performing an expansion of the energy in the high-temperature limit, and such a calculation yields

$$U = \frac{3}{2} N k_B T \left[ 1 - \frac{\zeta\left(\frac{3}{2}\right)}{2^{\frac{5}{2}}} \left(\frac{T_0}{T}\right)^{\frac{3}{2}} + \cdots \right],$$

(9.63)

which implies

$$C_V = \frac{3}{2} N k_B \left[ 1 + 0.231 \left(\frac{T_0}{T}\right)^{\frac{3}{2}} + \cdots \right],$$

(9.64)

with higher-order terms entering as integer powers of $(T_0/T)^{\frac{3}{2}}$. The first term in the expression is the classical heat capacity for an ideal gas, and then the remaining terms are quantum corrections.

For temperatures $T < T_0$, the derivative of $C_V$ with respect to $T$ is positive, whereas for temperatures $T > T_0$, the derivative of $C_V$ with respect to $T$ is negative. There is a discontinuity in $dC_V/dT$ at $T = T_0$, and a cusp in the heat capacity of an ideal Bose gas as illustrated in Fig. 9.9 (with the ideal gas result for comparison).[4] In $^4$He the heat capacity does not match very closely with the ideal Bose gas form we calculated above, however, there is a large peak in the heat capacity at the temperature at which helium becomes superfluid, known as the lambda peak, as shown in Fig. 9.10, which is reminiscent of the cusp for the non-interacting Bose gas.

[4] The careful reader will notice that while the expression for $C_V$ for $T < T_0$ is exact, the expression for $T > T_0$ is only approximate and might be concerned that additional terms in the perturbative expansion might smooth out the cusp. It can be shown that these additional terms do not in fact smooth out the cusp in the heat capacity. Full details of the calculation of $C_V$ are presented in Appendix C.

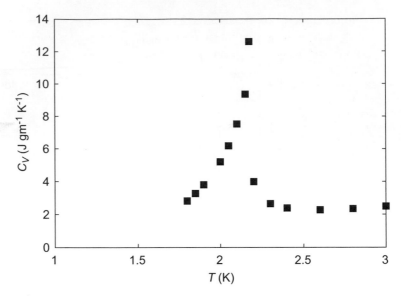

**Fig. 9.10** Specific heat of $^4$He at temperatures in the vicinity of the lambda peak at 2.176 K. Data points taken from Hill and Lounasmaa (1957).

The presence of discontinuities in thermodynamic quantities such as the derivative of the heat capacity is a feature of phase transitions. We will discuss this issue further in Chapter 10.

## 9.4 Bosonic Excitations: Phonons and Magnons

We saw earlier that photons can be described as massless bosons with zero chemical potential. There are several other contexts where bosons with $\mu = 0$ can arise in materials: excitations of the lattice (phonons) or of spin degrees of freedom (magnons). Phonons and magnons arise in systems with interactions but can behave like non-interacting bosons – they are basically normal modes of lattice and magnetic degrees of freedom of the system. We have already seen an example of such behaviour in the case of superfluid $^4$He, where the low-energy excitations of a strongly interacting system are phonons that behave like non-interacting bosons.

### 9.4.1 Phonons

Many properties of solids such as heat capacity or electrical resistance are strongly influenced by the excitations of vibrations in the solid (elastic deformations of a crystal lattice), which may be treated as sound-like waves. The quanta of these vibrations are phonons, similar to photons, which are quanta of light. Atoms in a solid interact quite strongly with each other, so calculating the vibrational properties of a solid containing on the order of $10^{23}$ atoms seems to be quite a daunting proposition. However, if the vibrations are small in amplitude, we can approximate the interaction potentials between atoms as harmonic, leading to the energy corresponding to the $j^{\text{th}}$ oscillator as

$$\epsilon_j = \hbar\omega_j \left(n_j + \frac{1}{2}\right),\tag{9.65}$$

where $n_j$ corresponds to the number of quanta (i.e. phonons) in the mode with frequency $\omega_j$. Phonons are quanta of simple harmonic oscillators, just like photons, and hence – as we saw previously in Eq. (4.49), the average number of phonons in a mode with frequency $\omega$ is given by the Bose–Einstein distribution with $\mu = 0$:

$$\langle n \rangle = \frac{1}{e^{\beta\hbar\omega} - 1}.\tag{9.66}$$

There is a lower limit to the allowed wavelength for phonons, which is given by the lattice spacing in the crystal $a$ – this is the smallest length scale on which there can be oscillations of the positions of the atoms in a crystal. There are three allowed polarizations of phonons (two transverse and one longitudinal), so in a crystal with $N$ atoms, there are $3N$ phonon modes. (One can obtain the same result by viewing each atom as a three-dimensional harmonic oscillator, and with $N$ atoms, there will be $3N$ modes in total.)

### 9.4.2 The Debye Model

The Debye model captures most of the important features of phonons, while mainly ignoring material-specific details. We ignore any differences in sound speed between polarizations and treat all modes as having a sound speed $s$. The existence of a finite number of phonon modes implies that there must be some maximum frequency, say $\omega_D$, since

$$N_{\text{phonon}} = 3N = \int_0^{\hbar\omega_D} d\epsilon\, g(\epsilon),\tag{9.67}$$

where $g(\epsilon)$ is the phonon density of states. We now switch from energy $\epsilon = \hbar\omega$ to angular frequency $\omega$. At low frequencies, it is usually a good approximation that $\omega = sk$, where $s$ is the speed of sound in a material. At higher frequencies, this is not correct due to anharmonicities in the potential. However, we can learn a lot from a model that assumes that $\omega = sk$ up to a cutoff frequency $\omega_D$. This model is known as the Debye model, and $\omega_D$ as the Debye frequency.

The phonon dispersion in the Debye model is the same as for photons, with $s$ replacing $c$, hence we can recall Eq. (9.9) and write down the density of states (recalling there are three polarizations rather than two, so the degeneracy factor is different) as

$$g(\omega) = \frac{3V\omega^2}{2\pi^2 s^3},\tag{9.68}$$

and hence

$$3N = \int_0^{\omega_D} d\omega\, g(\omega) = \frac{3V}{2\pi^2 s^3}\int_0^{\omega_D} d\omega\, \omega^2 = \frac{V\omega_D^3}{2\pi^2 s^3},\tag{9.69}$$

which we can invert to determine

$$\omega_D = \left(6\pi^2 n\right)^{\frac{1}{3}} s, \tag{9.70}$$

where $n = N/V$ is the number density of atoms in the material. The only place that material-specific quantities enter the Debye model is through the number density $n$ and the sound speed $s$.

We can calculate the mean energy of the phonons in the Debye model as

$$
\begin{aligned}
U &= \int_0^{\omega_D} d\omega\, g(\omega)\, \hbar\omega\, \langle n \rangle \\
&= \frac{3V\hbar}{2\pi^2 s^3} \int_0^{\omega_D} d\omega \frac{\omega^3}{e^{\beta\hbar\omega} - 1} \\
&= \frac{3V}{2\pi^2 s^3} \frac{(k_B T)^4}{\hbar^3} \int_0^{x_D} dx \frac{x^3}{e^x - 1},
\end{aligned} \tag{9.71}
$$

where we made the integral over $\omega$ dimensionless by changing variables to $x = \hbar\omega/k_B T$ and introduced

$$x_D = \frac{\hbar\omega_D}{k_B T} = \frac{\hbar s}{k_B T}\left(6\pi^2 n\right)^{\frac{1}{3}} = \frac{T_D}{T}, \tag{9.72}$$

with $T_D$ the Debye temperature. Rewriting our expression for $U$ in terms of $T_D$:

$$U = 9Nk_B T \left(\frac{T}{T_D}\right)^3 \int_0^{T_D/T} dx \frac{x^3}{e^x - 1}, \tag{9.73}$$

we notice that this is very similar to the integral we performed to find the energy of the photon gas, with the distinction that the upper cutoff is finite and temperature dependent rather than infinite. This means that we can no longer obtain a closed-form solution for all temperatures.

At very low temperatures, where $T \ll T_D$, we can see that the upper limit of the integral should be able to be replaced by $\infty$ with negligible change in the result, and so (using Eq. (9.13) with $m = 3$)

$$
\begin{aligned}
U &= 9Nk_B T \left(\frac{T}{T_D}\right)^3 \zeta(4)\Gamma(4) \\
&= \frac{3\pi^4 Nk_B T^4}{5 T_D^3},
\end{aligned} \tag{9.74}
$$

and hence the heat capacity takes the value

$$C_V = \frac{12\pi^4 Nk_B}{5}\left(\frac{T}{T_D}\right)^3, \tag{9.75}$$

which is known as the Debye $T^3$ law. This correctly captures the behaviour of the low-temperature heat capacity of many materials. Typical values for $T_D$ are on the order of 100–400 K.

At high temperatures, we can calculate the energy and hence the heat capacity by noting that the upper limit of the dimensionless integral goes to zero as $T \to \infty$, in which case we can approximate the integrand by the form it takes near $x = 0$, i.e.

$$\int_0^{T_D/T} dx \frac{x^3}{e^x - 1} \simeq \int_0^{T_D/T} dx \frac{x^3}{1 + x + \cdots - 1} \simeq \int_0^{T_D/T} dx\, x^2 = \frac{1}{3}\left(\frac{T_D}{T}\right)^3, \qquad (9.76)$$

and hence

$$U \simeq 3Nk_BT, \qquad (9.77)$$

which implies

$$C_V = 3Nk_B, \qquad (9.78)$$

which is exactly the Dulong and Petit law that we obtained earlier in Eq. (4.137) from the equipartition theorem. The full temperature dependence of the heat capacity in the Debye model is displayed in Fig. 9.11, demonstrating the regions of validity of the Debye $T^3$ law and the Dulong and Petit law and a comparison to experimental data for Al.

Despite the highly non-realistic dispersion assumed by the Debye theory, it gives an excellent account of the heat capacity of many materials, e.g. Al, Cu, Pb, not only at low temperatures and high temperatures, but also at intermediate temperatures. In particular, it matches the observation that $C_V \sim T^3$ at low temperatures and is independent of temperature at high temperatures, even though the Debye density of states generally does not match the true density of states over much of the phonon frequency range, as illustrated in the example of Al in Fig. 9.12.

Why does the Debye model work so well? The reason for this is that the model captures the important physics in the high- and low-temperature regimes. In the low-temperature regime, only the low-energy (low-frequency) density of states matters because these are the only occupied phonon modes, and the Debye model captures this correctly. At high

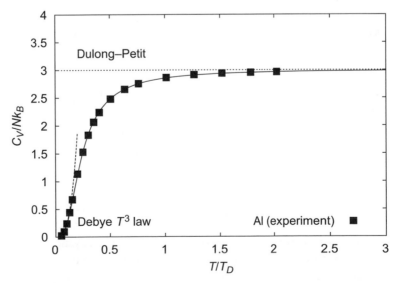

**Fig. 9.11** Heat capacity as a function of temperature in the Debye model. The low-temperature $T^3$ approximation and high-temperature Dulong–Petit limit are shown along with experimental data for Al taking $T_D = 395$ K (Stedman *et al.*, 1967).

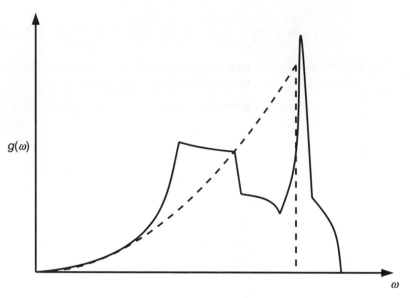

$g(\omega)$

**Fig. 9.12** Schematic comparison of the phonon density of states $g(\omega)$ as a function of angular frequency $\omega$ for Al (solid line) and the Debye model (dashed line).

temperatures, the heat capacity can be determined from the equipartition theorem, so the fact that the Debye model contains the correct number of modes ensures it gets this physics correct. Since both low- and high-temperature regimes are accurately represented, and the model interpolates smoothly between them, it is perhaps not so surprising that it gets the temperature dependence of $C_V$ correct over a wider temperature range, since a smooth interpolation is actually fairly well constrained if there are no structural phase transitions between the two limits. The Debye model also works well because of the presence of the Bose–Einstein distribution. At low temperatures, the Bose–Einstein distribution decays very quickly with $\omega$ so it heavily weights small $\omega$, which is where the density of states is most accurate.

### 9.4.3 Magnons

The spin waves found in ferromagnets share the property of vibrational modes of a crystal in that they can be treated as bosons with $\mu = 0$. In the ground state of a ferromagnet, all spins are parallel.[5] The low-energy excitations of a ferromagnet are spin waves, where the spin orientation varies slightly from site to site, and the spins precess about the overall magnetization direction, as illustrated in Fig. 9.13. A quantum treatment of the problem shows that spin waves have spin-1 and are hence bosonic.

Spin waves can be created or destroyed, just like phonons and photons, but they differ in that in many systems their low-energy dispersion is quadratic, rather than linear in $k$, so $\epsilon_k = \alpha k^2$. This leads to a different density of states compared to phonons, and hence the

---

[5] We will discuss ferromagnetism in detail in Chapter 10.

**Fig. 9.13** One wavelength of a spin wave in a ferromagnet. The spins precess around the overall magnetization direction in a ferromagnet. The longer the wavelength of a spin wave, the lower the energy.

low-temperature behaviour of the energy in three dimensions is

$$U = \int_0^\infty d\epsilon\, g(\epsilon)\, \epsilon \frac{1}{e^{\beta\epsilon} - 1}$$

$$\propto (k_B T)^{\frac{5}{2}} \int_0^\infty dx \frac{x^{\frac{3}{2}}}{e^x - 1}, \tag{9.79}$$

from which we see that at low temperatures the magnon contribution to the heat capacity should be proportional to $T^{\frac{3}{2}}$. This result follows from knowing the dimensionality of the system, the dispersion of the magnons and that they obey Bose statistics.

From considering fermionic and bosonic models, we have seen a wide variety of temperature dependences for heat capacity at low temperatures. Measurements of the temperature dependence of the heat capacity of materials at low temperatures can hence give clues as to the physical processes that are important at low temperatures in those materials.

## 9.5 Summary

In Bose systems, the occupation numbers are governed by the Bose–Einstein distribution

$$n(\epsilon) = \frac{1}{e^{\beta(\epsilon-\mu)} - 1}. \tag{9.80}$$

The quanta of the electromagnetic field, photons, can be described as bosons with chemical potential $\mu = 0$ and this has several important implications for their statistical mechanics. The energy density of thermal photons (blackbody radiation) takes the form found by Planck as a function of frequency:

$$u(\omega) = \frac{\hbar}{\pi^2 c^3} \frac{\omega^3}{e^{\beta\hbar\omega} - 1}, \tag{9.81}$$

and the power radiated per unit area in blackbody radiation is given by the Stefan–Boltzmann law

$$J = \sigma_{SB} T^4, \tag{9.82}$$

where $\sigma_{SB}$ is the Stefan–Boltzmann constant.

For massive non-interacting bosons, as the temperature is lowered towards zero, the chemical potential approaches zero from below and the phenomenon of Bose–Einstein condensation occurs, in which there is macroscopic occupation of the lowest-energy orbital. If there are interactions between the bosons, such as in $^4$He, a condensate can still form, and it gives rise to superfluidity, where the fluid can flow without resistance.

Excitations of lattice and magnetic degrees of freedom in solids, phonons and magnons, respectively can also be described as bosons with chemical potential $\mu = 0$. The Debye model, which takes into account the finite number of phonon modes and approximates the phonon density of states as having the form for sound waves, can give a very good account of the heat capacity of a variety of materials.

# Problems

**9.1**  Show that the grand potential for an ideal Bose gas for temperatures $T < T_0$ takes the form

$$\Phi = -\zeta \left(\frac{5}{2}\right) k_B T n_Q V - k_B T \ln \left(1 - e^{\beta \mu}\right).$$

By using Eq. (9.55), show that the second term can be simplified to a form which is manifestly subextensive. Use this result to show that the entropy of the gas for $T < T_0$ in the thermodynamic limit is

$$S = \frac{5}{2} \zeta \left(\frac{5}{2}\right) k_B n_Q V.$$

By comparing this expression for the number density of bosons in the normal state, deduce that the entropy per particle in the normal fluid is

$$\frac{5}{2} k_B \frac{\zeta \left(\frac{5}{2}\right)}{\zeta \left(\frac{3}{2}\right)}.$$

**9.2**  *Luminosity of the Sun*
The surface temperature of the Sun is $T \sim 5800$ K, and its radius is $R_\odot \sim 6.96 \times 10^5$ km. Calculate its total power output (i.e. its luminosity). From this result calculate the total energy from the Sun incident on the Earth every day. The Earth is at a distance of $149.6 \times 10^6$ km from the Sun and has an average radius of 6371 km.

**9.3**  The idea of a solar sail is to use the momentum carried by photons emitted by the Sun to propel a spacecraft.

(a) Use the Stefan-Boltzmann formula to estimate the incident power per unit area at the Earth's orbit from the Sun.
(b) Given your answer to part (a), estimate the force on an 800 m × 800 m reflecting solar sail located at the same distance from the Sun as the Earth due to photons from the Sun.
(c) Estimate the change in velocity for a 5000 kg spaceship with the solar sail in part (b) attached after the first day of operating the sail.

**9.4** For each type of particle below, give the dependence of the chemical potential $\mu$ on the density $n$, using calculations where appropriate:

(a) Maxwell–Boltzmann particles in three dimensions.
(b) Photons.
(c) Non-relativistic fermions in three dimensions at zero temperature.
(d) Non-relativistic bosons at low but non-zero temperature.

Illustrate your answer with a graph showing $\mu$ against $n$ for cases (a) to (d).

**9.5** (a) Show that the grand canonical partition function for blackbody radiation takes the form

$$\Xi = \prod_s \frac{1}{1 - e^{-\beta\epsilon_s}}.$$

(b) Using part (a) or otherwise, show that the entropy per unit volume of blackbody radiation at temperature $T$ is

$$s = \frac{4\pi^2 k_B^4 T^3}{45\hbar^3 c^3}.$$

**9.6** Show that for non-interacting bosons with a dispersion $\epsilon = \hbar s k$, where $s$ is a velocity, the average occupation of the lowest-energy orbital is

$$\frac{\langle N_0 \rangle}{N} = 1 - \left(\frac{T}{T_0}\right)^3,$$

and identify $T_0$. Note the difference in temperature dependence from Eq. (9.39).

**9.7** Consider a non-relativistic ideal Bose gas composed of particles which have an internal degree of freedom. The internal levels are $\epsilon_0 = 0$ and $\epsilon_1 > \epsilon_0$. Determine an equation for the Bose–Einstein condensation temperature $T_0$ of this gas in three dimensions as a function of $\epsilon_1$. (You do not need to solve this equation.)

**9.8** Bose–Einstein condensation of dilute gas of $^{87}$Rb atoms was first reported in 1995 (Anderson *et al.*, 1995).

(a) Is $^{87}$Rb (37 protons and electrons, 50 neutrons) a boson or a fermion?
(b) At their quoted maximum number density of $n = 2.5 \times 10^{12}$ cm$^{-3}$, at what temperature $T_c^{\text{predict}}$ do you expect the onset of Bose condensation in free space? Their claim is that they found Bose condensation at a temperature $T_c^{\text{measured}}$ of 170 nK. How does this compare to your estimate? (Anderson *et al.* considered bosons confined in a trap, so you should not necessarily expect the numbers to agree.)
(c) A more quantitative estimate than part (b) can be obtained by approximating the trapping potential by a three-dimensional harmonic oscillator potential

$$V(\mathbf{r}) = \frac{1}{2}m\omega^2(x^2 + y^2 + z^2),$$

for which the energies $\epsilon_{n_x,n_y,n_z}$ of the orbitals are labelled by the quantum numbers $n_x$, $n_y$, $n_z$ and

$$\epsilon_{n_x,n_y,n_z} = \hbar\omega \left( n_x + n_y + n_z + \frac{3}{2} \right).$$

Show that for large $\epsilon$ the density of states is given by

$$g(\epsilon) = \frac{\epsilon^2}{2(\hbar\omega)^3}.$$

(d) Use your result from part (c) to obtain an estimate for $T_c$ for $N$ atoms in the trap as a function of $N$ and $\omega$ when $N$ is large.

**9.9** Following a similar approach to the calculation we considered to find the temperature $T_0$ for Bose–Einstein condensation in three dimensions, calculate the critical temperature for Bose–Einstein condensation in two dimensions. Do not be surprised if your result differs significantly from the three-dimensional case.

*Hint:* find the chemical potential as a function of temperature.

**9.10** Consider a three-dimensional ideal Bose gas with energy spectrum $\epsilon_k = Ak^\nu$.

(a) Assuming that Bose–Einstein condensation occurs, find the Bose–Einstein condensation temperature $T_0$.
(b) For what values of $\nu$ do you expect Bose–Einstein condensation to occur?

**9.11** Show that for non-relativistic ideal bosons the average pressure $P$ is related to the average energy $U$ by

$$P = \frac{2U}{3V}.$$

**9.12** Show that for ultra-relativistic ideal bosons (for which the dispersion is $\epsilon = pc$) the average pressure $P$ is related to the average energy $U$ by

$$P = \frac{U}{3V}.$$

**9.13** At the centre of the Sun the temperature is $T = 1.6 \times 10^7$ K and the plasma there is composed of hydrogen with density $\rho_H = 6 \times 10^4$ kg m$^{-3}$ and helium with density $\rho_{He} = 1 \times 10^5$ kg m$^{-3}$.

(a) Calculate the ratio $n/n_Q$ for the electrons, protons and He nuclei at the centre of the Sun.
(b) Estimate the pressure at the centre of the Sun from these particles and also that from radiation pressure and determine whether it is pressure due to particles or radiation that prevents the gravitational collapse of the Sun.

**9.14** Consider the Coulomb potential between two ions in a crystal and show that this gives rise to a harmonic potential for small displacements from their equilibrium positions.

**9.15** Consider a dielectric crystal consisting of layers of atoms with rigid coupling between layers so that the motion of atoms is restricted to the plane of the layer. Show that the phonon heat capacity in the Debye approximation in the low-temperature limit is proportional to $T^2$.

**9.16** Find the first few terms in the low-temperature expansion of the heat capacity from magnons that have a dispersion of the form $\epsilon_k = \Delta + \alpha k^2$.

# 10      Phase Transitions and Order

We are all familiar with different phases of matter, such as solid, liquid and gas, and some phase transitions such as solid to liquid (e.g. ice melting) or liquid to gas (e.g. water boiling). Phases of matter that we have touched on, or will discuss in more detail, include superfluids, superconductors and magnetic phases such as ferromagnetism (for example, when a lump of iron is cooled below 1043 K it develops a spontaneous permanent magnetization). Other phases or phase transitions that we will not discuss in detail here, but are of considerable interest, include liquid crystals, the many different types of crystal lattice and metal–insulator transitions. Many such phases and phase transitions can be understood within a framework where symmetry defines the order in a particular phase, and the appearance or disappearance of order is associated with a phase transition. We will mainly discuss the Ising model, which is a simplified but very instructive model that contains important aspects of the physics of ferromagnetism. Many of the concepts that we develop in this setting, and generalizations of the Ising model, have much wider applicability in magnetism, the binding of molecules to surfaces, neuroscience and glassy systems.

## 10.1   Introduction to the Ising Model

We discussed the behaviour of non-interacting spins in a magnetic field in Chapters 1 and 2. We saw that the spins align in the strong-field, low-temperature limit. In general there can also be interactions between the spins, in addition to the magnetic field. An obvious origin for these interactions is the magnetic dipole–dipole interaction between two atoms. These interactions give a temperature scale of the order of 1 K, which is far too small to be able to explain phenomena such as magnetic ordering in iron at a temperature of over 1000 K.[1] The usual origin of the interactions is electron exchange. This arises because of the Pauli exclusion principle. The spatial part of the wavefunction of a pair of electrons has a different extent, depending on whether the electrons are in the same spin state or not, because they cannot be in the same location and have all their other quantum numbers be identical if they have the same spin, due to the exclusion principle. On the contrary, there is no restriction if they have opposite spins. This spatial difference in wavefunctions, combined with the Coulomb interaction between the electrons, leads to an energy that depends on whether the two electron spins are

---

[1] Dipole interactions can be important in some materials at low temperatures.

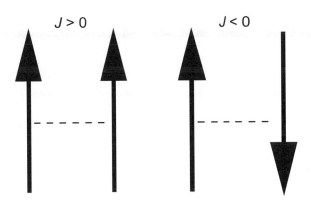

**Fig. 10.1** Comparison of the lowest-energy spin configurations for a pair of spins with $J > 0$ and $J < 0$.

aligned or not. This corresponds to an effective interaction between spins, where the energy depends on their relative orientation.

The full treatment of the statistical mechanics of many quantum mechanical spins can be quite challenging, as the spin operators corresponding to different spin components do not commute. We will not concern ourselves with quantum mechanical spins and will only focus on classical representations of spins in which a spin is represented as a vector in space.

The simplest form of interaction is a pairwise one, so that for two interacting vector spins $\mathbf{s}_1$ and $\mathbf{s}_2$, the energy of the configuration is given by

$$
\begin{aligned}
E &= -J\mathbf{s}_1 \cdot \mathbf{s}_2 \\
&= -J|\mathbf{s}_1||\mathbf{s}_2| \cos \theta,
\end{aligned}
\tag{10.1}
$$

where $\theta$ is the angle between the spins and $J$ is the exchange energy (the details of the wavefunction and Coulomb interaction are buried in this term – we will treat it as a parameter and not worry about its origin further). The configuration that the pair of spins choose will depend on the sign of $J$. If $J > 0$, the lowest energy is obtained for $\cos \theta = 1$, i.e. $\theta = 0$, in which case the spins line up parallel to each other. If $J < 0$, the lowest-energy configuration has $\cos \theta = -1$, i.e. $\theta = \pi$, and so the two spins align antiparallel to each other (Fig. 10.1).

Assuming that pairwise interactions are the most important class of interaction (in many cases this is true), we can generalize from the two-spin example to many spins sitting on a lattice, with the spin on lattice site $i$ denoted by $\mathbf{s}_i$. The interaction energy is then

$$
E = -J \sum_{\text{pairs}, i \neq j} \mathbf{s}_i \cdot \mathbf{s}_j,
\tag{10.2}
$$

which is known as the Heisenberg model.

We already made the simplifying assumption that we can ignore the quantum mechanical origin of the spins, and consider them as classical vectors.[2] Even more drastically, we will

---

[2] Solving the problem specified by Eq. (10.2) when the spins are quantum mechanical is much more difficult than the classical case and a particularly active area of research for $J < 0$.

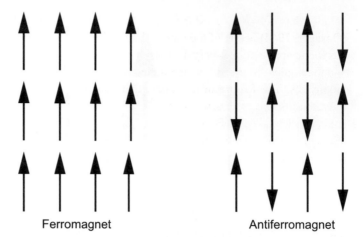

Ferromagnet                              Antiferromagnet

**Fig. 10.2**        Ferromagnetic and antiferromagnetic spin configurations on a square lattice.

only allow the spins to point up or down and only interact with their nearest neighbours (n.n.),[3] which gives the most commonly studied version of the Ising model:[4]

$$E = -J \sum_{\langle i,j \rangle} s_i s_j, \tag{10.3}$$

where the notation $\langle i, j \rangle$ indicates to sum only over n.n. pairs of sites $i$ and $j$. The spins $s_i$ and $s_j$ can only take on the values $\pm 1$ and are known as *Ising spins*.[5]

The sign of $J$ again determines the type of ordering that results from the expression for the energy, Eq. (10.3). If $J > 0$, then the lowest-energy configuration corresponds to all spins pointing in the same direction, which is what happens in a ferromagnet (note that if all spins point up this has the same energy as when all spins point down). Alternatively, if $J < 0$, then the lowest-energy configuration (at least on a square or cubic lattice) is antiferromagnetic, consisting of spins that are antiparallel on neighbouring sites if we restrict the summation to neighbouring spins only.[6] The spin configurations in a ferromagnet and an antiferromagnet are illustrated schematically in Fig. 10.2.

The Ising model is the simplest statistical mechanical model of interacting spins on a lattice. Despite its apparent simplicity, it is not particularly easy to solve. In fact, no analytic

---

[3] The assumption of n.n. interactions is usually a good approximation, as the strength of interactions generally decreases quickly with increasing separation between spins. However, there are materials where it is important to include some longer-range spin interactions.

[4] The Ising model was not actually developed by Ising: it was created by Ernst Ising's PhD supervisor, Wilhelm Lenz, and given as a problem to Ising. Ising found the solution of the model in one dimension and published the result in 1925.

[5] The approximation to Ising spins might seem rather artificial, but in general exchange interactions can vary depending on the orientation of spins due to their crystalline environment, i.e. a more general spin interaction than $J\mathbf{s}_1 \cdot \mathbf{s}_2$ would be $\sum_{\alpha,\beta=x,y,z} J_{\alpha\beta} s_1^\alpha s_2^\beta$. An anistropic crystalline environment can lead to a situation in which Ising spins are a good representation of the allowed magnetic degrees of freedom.

[6] Flipping the sign of $J$ from positive to negative changes a ferromagnetic ground state to an antiferromagnetic one if the lattice is bipartite, i.e. we can define distinct $A$ and $B$ sublattices. Square and cubic lattices are examples of bipartite lattices, whereas triangular lattices are not.

solution has been found in three dimensions, and Onsager's solution in two dimensions (Onsager, 1944) is one of the more celebrated results in statistical mechanics. However, it has found application in very diverse physical systems, including magnetism, binary alloys and the liquid–gas transition, and in fields such as neural networks, information theory and biophysics. The feature of the model that makes it much more interesting than the non-interacting spin systems we studied in Chapters 1 and 2 is that the interactions between spins can lead to collective behaviour that is not possible in the non-interacting system. We will consider the simplest version of the model in which all spins interact with the same exchange interaction $J$. One can have more general variants in which the exchange interactions can vary for each pair of sites $i$ and $j$.

## 10.2 Solution of the Ising Model

We now specify carefully the model that we will study. We consider a lattice of $N$ sites, where at each site $i$ there is a spin $s_i$ that can take one of the values $\pm 1$. The energy for a given spin configuration $\{s_i\}$ is

$$E(\{s_i\}) = -J \sum_{\langle i,j \rangle} s_i s_j - H \sum_i s_i, \tag{10.4}$$

where usually the sum over $i$ and $j$ is restricted to sites $i$ and $j$ that are nearest neighbours. The parameter $H$ represents an external magnetic field experienced by the spins, and $J$ is the exchange energy associated with spins $s_i$ and $s_j$. In the absence of a field ($H = 0$), if $J > 0$, then the energy associated with a pair of spins is $-J s_i s_j$, and this is minimized if $s_i = s_j$. At $T = 0$, the spins will all be aligned and either all point up or all point down, forming a *ferromagnetic* phase. If $J < 0$, then the energy is minimized if $s_i = -s_j$, so that neighbouring spins are antiparallel. Such a phase is termed an *antiferromagnet* and illustrated in Fig. 10.2. We will now focus solely on the situation in which $J > 0$, unless otherwise specified.

We know that at equilibrium the Helmholtz free energy $F = U - TS$ is minimized – we can use this information to obtain the expected behaviour of the model in the very low- and very high-temperature limits. At $T = 0$, the energy considerations we have given above determine the nature of the arrangement of spins in their ground state, as shown for a square lattice in Fig. 10.2. At high temperatures, the free energy is minimized by maximizing the entropy, because $U$ is bounded. The maximum entropy is achieved when all $2^N$ microstates are equally likely, giving an entropy of $S = N k_B \ln 2$. If all microstates are equally likely, then the magnetization will be determined by those which are most numerous, and these correspond to those with equal (or very close to equal) numbers of up and down spins so there is no net magnetization in the *paramagnetic* phase.

### 10.2.1 Order Parameters and Broken Symmetry

Our arguments above suggest that there should be distinct high- and low-temperature behaviour of the Ising model in the absence of a magnetic field ($H = 0$): a high-temperature phase with random spin orientations, and a low-temperature phase with ferromagnetic order. One way to characterize these phases is by using the magnetization $M = \langle \sum_i s_i \rangle$. In the low-temperature phase, the magnetization per spin $m = M/N$ is non-zero and approaches 1 as $T \to 0$. In the paramagnetic phase, the magnetization per spin $m = M/N$ is zero (in the limit $N \to \infty$). The magnetization $m$ is often referred to as an *order parameter* for the ferromagnet, since it is non-zero only in the ordered phase, and is zero in the disordered phase (the paramagnet). The temperature at which $m$ becomes non-zero is where ferromagnetic order first appears, and is called the Curie temperature. The idea of order parameters is much more widespread than just magnetic examples, and in general the temperature at which the order parameter becomes non-zero is known as the *critical temperature*.

The model specified in Eq. (10.4) has a symmetry when $H = 0$ that the energy is unchanged if we flip the sign of every spin.[7] However, as we have argued above, as $T \to 0$, the state of the system is one in which there is a non-zero magnetization, which means that a particular orientation of the spins has been chosen, breaking the up–down symmetry present at high temperatures in the paramagnetic phase where $m = 0$. In general, the development of an order parameter coincides with the breaking of some symmetry of the system. Another example of broken symmetry that should be somewhat familiar is a crystal – the continuous translational symmetry of a liquid is broken to a discrete translational symmetry in the crystal – the system is only invariant under translation by a lattice vector.

### 10.2.2 General Strategy for Solution of the Ising Model

In general, the partition function of the Ising model is

$$Z = \sum_{\{s_i\}} e^{-\beta E(\{s_i\})}, \tag{10.5}$$

where we sum over all microstates, i.e. spin configurations $\{s_i\} = \{s_1, s_2, \ldots, s_N\}$, and $E(\{s_i\})$ is the energy associated with the spin configuration $\{s_i\}$. The hard part of the problem is evaluating the partition function – in one dimension this can be done exactly, and in two dimensions it can be done for $H = 0$, but for two dimensions with $H \neq 0$ or in three dimensions, there is no closed-form solution. Once we have a form (exact or approximate) for $Z$, then we can determine the Helmholtz free energy $F = -k_B T \ln Z$. With knowledge of the Helmholtz free energy, we can determine the magnetization via

$$M = \left\langle \sum_i s_i \right\rangle$$

$$= \frac{1}{Z} \sum_{\{s_i\}} \left[ \left( \sum_i s_i \right) e^{-\beta \left[ -J \sum_{\langle i,j \rangle} s_i s_j - H \sum_i s_i \right]} \right]$$

---

[7] A more general example is provided by Eq. (10.2), which has the symmetry property that the energy is unaffected if all spins are rotated by the same fixed amount.

$$
\begin{aligned}
&= \frac{1}{Z} \frac{1}{\beta} \frac{\partial Z}{\partial H} \\
&= \frac{k_B T}{Z} \frac{\partial Z}{\partial H} \\
&= \frac{\partial}{\partial H} \left( k_B T \ln Z \right) \\
&= -\frac{\partial F}{\partial H}.
\end{aligned}
\tag{10.6}
$$

Additionally, the magnetic susceptibility can be determined from

$$
\chi_M = \frac{\partial M}{\partial H},
\tag{10.7}
$$

and is a measure of the strength of ferromagnetic fluctuations in the system. The magnetic susceptibility tells us how the magnetization changes due to a small change in field. It will be largest when $M$ is close to zero, but can be tipped to a different value with a small change in field, and hence provides a good way to identify a phase transition – as we will see below.

We first introduce mean field theory as a means to obtain an approximate solution of the Ising model and then show how to solve the Ising model exactly in one dimension.

### 10.2.3  Mean Field Theory

We will develop an approach to calculating the partition function of the Ising model that captures much of the correct physics by making a mean field approximation. To do this, first focus on a single spin at site $i$. The contribution to the energy from this spin is

$$
E_i(s_i) = -s_i \left( H + \frac{J}{2} \sum_j s_j \right),
\tag{10.8}
$$

where the factor of $\frac{1}{2}$ is to ensure that

$$
\sum_i E_i = \sum_{\langle i,j \rangle} s_i s_j - H \sum_i s_i,
\tag{10.9}
$$

i.e. to ensure that we don't double count in the sum over $i$ and $j$.[8] We can view the energy $E_i$ as that of a spin in an effective magnetic field, one portion coming from the physical field $H$ and another from the interactions with neighbouring spins. We can also write (with no approximation)

$$
s_j = \langle s_j \rangle + \left( s_j - \langle s_j \rangle \right),
\tag{10.10}
$$

---

[8] The distinction between performing a summation $\sum_i \sum_j$ and a summation $\sum_{\langle i,j \rangle}$ is that in the first sum we are required to sum over every pair of neighbouring spins twice (once with each spin labelled as "$i$" and once with each spin labelled as "$j$"), whereas in the second sum we only sum over each pair of spins once. To take this into account, we shall sometimes need to multiply terms by factors of $\frac{1}{2}$, and will identify when this is the case, as we do in Eq. (10.8).

which implies that when we sum over the $z$ nearest neighbours to $s_i$:

$$\sum_j s_j = z\langle s_j \rangle + \sum_{j=1}^{z} \left( s_j - \langle s_j \rangle \right).$$  (10.11)

We can identify the two terms on the right-hand side of Eq. (10.11) as a mean field term, $z\langle s_j \rangle$, corresponding to the average spin for the whole lattice and a fluctuation term reflecting the local environment of that spin. Note that $z$ depends on dimensionality – for the example of a one-dimensional line, $z = 2$, for a two-dimensional square lattice, $z = 4$, and for a three-dimensional cubic lattice, $z = 6$. We can use the identity in Eq. (10.10) to rewrite the interaction term in the energy as

$$s_i s_j = \left[ \langle s_i \rangle + \left( s_i - \langle s_i \rangle \right) \right]\left[ \langle s_j \rangle + \left( s_j - \langle s_j \rangle \right) \right]$$

$$= s_i \langle s_j \rangle + \langle s_i \rangle s_j - \langle s_i \rangle\langle s_j \rangle + \left( s_i - \langle s_i \rangle \right)\left( s_j - \langle s_j \rangle \right).$$  (10.12)

Inserting Eq. (10.12) in Eq. (10.4) we find the energy for the Ising model to be

$$E = -\frac{J}{2}\sum_i \sum_{j \in \{n.n.\, i\}} \left[ s_i \langle s_j \rangle + \langle s_i \rangle s_j - \langle s_i \rangle\langle s_j \rangle \right] - H\sum_i s_i$$

$$+ J\sum_{\langle i,j \rangle}\left( s_i - \langle s_i \rangle \right)\left( s_j - \langle s_j \rangle \right),$$  (10.13)

where $j \in \{n.n.\, i\}$ indicates that $j$ is a nearest neighbour of $i$. The approximation we make is to drop the last term in Eq. (10.13), the sum over the product of fluctuations. This term will be small if

$$\sum_{j=1}^{z} s_j \simeq z\langle s_j \rangle,$$  (10.14)

i.e. every spin experiences an identical local environment on average. We know that each spin will generally have a different local environment from other spins in most microstates. However, if each spin has a large number of neighbours, then Eq. (10.14) becomes a reasonable approximation. In the limit that the number of neighbours goes to infinity, the approximation is exact, which we can anticipate from the central limit theorem (another way to state this is that mean field theory is exact in an infinite number of dimensions). However, this also points to the approximation being least accurate when the number of neighbours is small. We will see that it is in fact qualitatively incorrect in one dimension. When we also note that

$$\langle s_i \rangle = m = \frac{M}{N},$$  (10.15)

this allows us to simplify the expression for the energy to

$$E = \frac{zJm^2 N}{2} - zJm\sum_i s_i - H\sum_i s_i.$$  (10.16)

Our approximation to the partition function for $N$ spins is thus

$$Z_N = e^{-\frac{1}{2}\beta N z J m^2}\sum_{s_i = \pm 1} e^{z\beta Jm \sum_i s_i + \beta H \sum_i s_i}$$

$$= e^{-\frac{1}{2}\beta N z J m^2} \sum_{s_i=\pm 1} \prod_{i=1}^{N} e^{\beta(zJm+H)s_i}$$

$$= e^{-\frac{1}{2}\beta N z J m^2} \prod_{i=1}^{N} \sum_{s_i=\pm 1} e^{\beta(zJm+H)s_i}$$

$$= e^{-\frac{1}{2}\beta N z J m^2} \prod_{i=1}^{N} \{2\cosh[\beta(zJm+H)]\}$$

$$= e^{-\frac{1}{2}\beta N z J m^2} \{2\cosh[\beta(zJm+H)]\}^{N} . \qquad (10.17)$$

Apart from the exponential term, this partition function is exactly what we would have for $N$ distinguishable spins, each experiencing a field $zJm + H$. If we have a test spin $s_i$, we can view the magnetization of all of the remaining spins as providing an effective "mean field" that causes the spin $s_i$ to align with the background magnetization.

Equation (10.17) allows us to determine the mean field free energy as

$$F = -k_B T \ln Z_N$$

$$= \frac{N z J m^2}{2} - N k_B T \ln \{2\cosh[\beta(zJm+H)]\} . \qquad (10.18)$$

We can perform some simple sanity checks to see that this free energy behaves in accordance with our expectations at low and high temperatures. At low temperatures, if we let $T \to 0$, we get

$$F = \frac{N z J m^2}{2} - N(zJm+H)$$

$$= -\frac{N z J}{2} - NH$$

$$= U(T = 0) \qquad (10.19)$$

(provided $m(T = 0) = 1$). In the high-temperature limit as $T \to \infty$ we get

$$F = -N k_B T \ln 2$$

$$= -TS(T = \infty) \qquad (10.20)$$

(provided $m \to 0$) where, as we already noted, the maximum entropy is $S(T = \infty) = N k_B \ln 2$. From a knowledge of the free energy we can determine the magnetization:

$$M = -\frac{\partial F}{\partial H} = N \tanh(\beta(zJm+H)) . \qquad (10.21)$$

If we set $H = 0$ and recall that $m = M/N$, then we obtain a transcendental equation for the magnetization:

$$m = \tanh(z\beta J m). \qquad (10.22)$$

This can be solved graphically (as illustrated in Fig. 10.3) or numerically. At high temperatures, i.e. those such that $z\beta J < 1$, the only solution is $m = 0$. However, at lower temperatures when $z\beta J \geq 1$, in addition to the solution at $m = 0$, there is also a solution with non-zero $m$. We can see from the expression for the free energy, Eq. (10.18), that

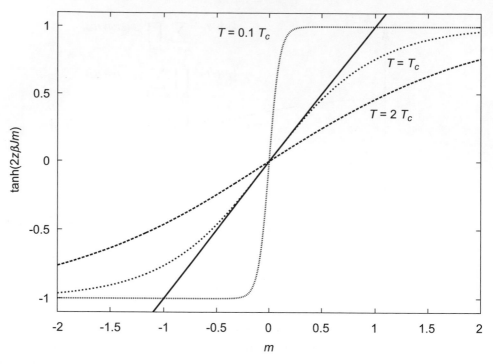

Fig. 10.3 Graphical solution of Eq. (10.22) for magnetization at temperatures above, below and equal to $T_c$.

when $H = 0$, the solution with $m \neq 0$ can have a lower free energy than the solution with $m = 0$ when $z\beta J \gg 1$. This implies that at an ordering temperature of $z\beta_c J = 1$, or

$$T_c = \frac{zJ}{k_B}, \tag{10.23}$$

there is spontaneous ordering of the spins and a ferromagnetic phase appears. (The mean field theory overestimates the value of $T_c$ but does get the physics of ordering correct in dimensions 2 and larger.)

In the vicinity of the ordering temperature, $m$ is small, so we may expand the right-hand side of Eq. (10.22) to get

$$m = \tanh(z\beta Jm) \simeq \beta(zJm) - \frac{1}{3}\beta^3(zJm)^3, \tag{10.24}$$

so we have

$$m = \beta(zJm) - \frac{1}{3}\beta^3(zJm)^3, \tag{10.25}$$

which always has the solution $m = 0$, but also has a solution that can be obtained from solving

$$\frac{zJ}{k_B T} - 1 = \frac{1}{3}\left(\frac{zJ}{k_B T}\right)^2 m^2, \tag{10.26}$$

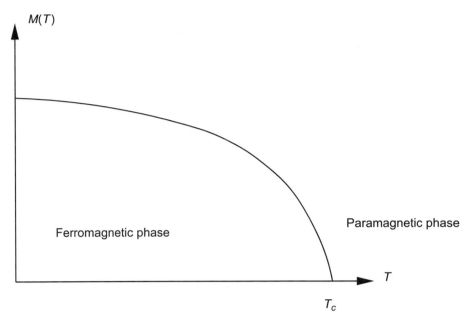

Fig. 10.4 Schematic of the magnetization in a ferromagnet as a function of temperature. The magnetization has a non-zero value in the ferromagnetic (low-temperature) phase and is zero in the paramagnetic (high-temperature) phase.

which we can rearrange to find that for $T < T_c$:

$$m(T) = \pm \left[ 3 \left( \frac{k_B T}{zJ} \right)^2 \left( \frac{zJ}{k_B T} - 1 \right) \right]^{\frac{1}{2}}$$

$$\simeq \pm \left[ 3 \frac{(T_c - T)}{T} \right]^{\frac{1}{2}}, \tag{10.27}$$

so we see that

$$m(T) \propto |T_c - T|^{\beta}, \tag{10.28}$$

where we noted that $T \simeq T_c = zJ/k_B$ and the exponent $\beta = 1/2$,[9] and that this form holds only for $T < T_c$ and $m \ll 1$. The behaviour of the magnetization as a function of temperature is illustrated schematically in Fig. 10.4.

From our expression for the magnetization as a function of temperature and field,

$$M = N \tanh \left[ \beta(zJm + H) \right], \tag{10.29}$$

we can also determine the magnetic susceptibility, recalling that $\chi_M = \partial M / \partial H$ and $m = M/N$. We will first evaluate $\chi_M$ and then set $H = 0$, because we are interested in studying what happens at the critical point. When we set $H = 0$ after taking the derivative

---

[9] We have generally used $\beta$ to indicate the inverse temperature $\beta = 1/k_B T$, but we also use it in Eq. (10.28) for the order parameter exponent for consistency with the literature on critical phenomena.

of Eq. (10.29) with respect to $H$, we have

$$\chi_M = N\left(\beta + zJ\beta\frac{\chi_M}{N}\right)\text{sech}^2(zJ\beta m),$$  (10.30)

which leads to

$$\chi_M\left[1 - zJ\beta\text{sech}^2(zJ\beta m)\right] = N\beta\text{sech}^2(zJ\beta m),$$  (10.31)

and hence

$$\chi_M = \frac{N\beta\text{sech}^2(zJ\beta m)}{\left[1 - zJ\beta\text{sech}^2(zJ\beta m)\right]}.$$  (10.32)

We can note that $zJ\beta = T_c/T$, and expand the expression for $\chi_M$ in the vicinity of $T_c$. We consider the cases of $T > T_c$ and $T < T_c$ separately.

## Magnetic susceptibility, $T > T_c$

If $T > T_c$, then $m = 0$, and we can note that $\text{sech}(0) = 1$ so that

$$\chi_M = \frac{N\beta}{\left[1 - \frac{T_c}{T}\right]}$$

$$= \frac{N}{k_B(T - T_c)},$$  (10.33)

so we see that the susceptibility has the form

$$\chi_M \propto |T - T_c|^\gamma,$$  (10.34)

with $\gamma = -1$.

## Magnetic susceptibility, $T < T_c$

For $T < T_c$, $m$ is non-zero and we need to make use of the result that we determined in Eq. (10.27), that

$$m \simeq \left(\frac{3}{T_c}\right)^{\frac{1}{2}}(T_c - T)^{\frac{1}{2}},$$  (10.35)

along with the expansion of $\text{sech}^2(x)$ for small $x$:

$$\text{sech}^2(x) = \frac{1}{\cosh^2(x)} \simeq \frac{1}{\left(1 + \frac{x^2}{2}\right)^2} \simeq 1 - x^2,$$  (10.36)

in which case, with $x = mT_c/T$,

$$\chi_M \simeq \frac{\beta N\left[1 - \left(\frac{T_c}{T}\right)^2\frac{3}{T_c}(T_c - T)\right]}{1 - \frac{T_c}{T}\left[1 - \frac{3}{T_c}(T_c - T)\right]}$$

$$\simeq -\frac{\beta N}{2\left(1 - \frac{T_c}{T}\right)},$$  (10.37)

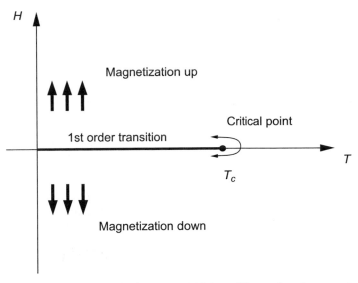

$H$

Magnetization up

↑↑↑

Critical point

1st order transition

$T_c$

$T$

↓↓↓

Magnetization down

Fig. 10.5 Magnetic field–temperature phase diagram for the Ising model. It is possible to go from the magnetization up region to the magnetization down region either by crossing a first-order transition for $T < T_c$, or by taking a trajectory in the $(H, T)$ plane that crosses the $H = 0$ line for $T > T_c$, in which case there is no phase transition. If $H = 0$, and $T$ is varied, then there is a second-order phase transition at the critical point $T = T_c$.

where we noted $T \simeq T_c$ and hence $(T_c - T)/T_c \ll 1$ in the numerator, but we kept all terms in the denominator so we get

$$\chi_M = -\frac{N}{2k_B(T - T_c)}$$
$$= \frac{N}{2k_B|T_c - T|}, \tag{10.38}$$

which is of the form

$$\chi_M \propto |T - T_c|^{\gamma'}, \tag{10.39}$$

where $\gamma' = \gamma = -1$. The susceptibility has the same form as above the transition, but is smaller by a factor of two. The divergence of the magnetic susceptibility at $T_c$ is a characteristic of a second-order phase transition such as the Ising transition. In a second-order phase transition, there is usually a discontinuity in the second derivative of the free energy (here the susceptibility), whereas in first-order transitions (such as ice melting) there is a discontinuity in the first derivative of the free energy (the latent heat associated with melting ice is an example of such a discontinuity).

The Ising model displays both first- and second-order transitions (Fig. 10.5). We have investigated the second-order transition when the temperature is varied at $H = 0$. It is also possible to have a first-order transition when $T < T_c$, in which $T$ is constant and the sign of $H$ is changed. If we change $H$ from negative to positive, then there is a discontinuous change in the order parameter, from a state with negative, non-zero magnetization to a state with positive, non-zero magnetization. This jump in the order parameter is a signature of a first-order phase transition.

## 10.3 Role of Dimensionality

The mean field solution to the Ising model we presented in Section 10.2.3 predicted ordering in any number of dimensions, although, as we discussed, it is strictly only correct in the infinite-dimensional limit, when $z \to \infty$. We will discuss the exact solution to the Ising model in one and two dimensions in Sections 10.4.1 and 10.4.2, respectively. Before doing so, we can give some simpler arguments to address whether we might expect to see ordering in each of these dimensions.

### 10.3.1 One Dimension

Consider the ground state of the one-dimensional Ising model, illustrated in Fig. 10.6. The exchange energy for each pair of spins is $-J$, so for a chain with $N + 1$ spins the ground-state energy is $E_0 = -NJ$ and the magnetization is $M = N + 1$. The lowest-energy excited state will be one in which one pair of spins is antiferromagnetically aligned, but all others remain ferromagnetically aligned, as illustrated in Fig. 10.7.

The microstates corresponding to this first excited state have a magnetic structure consisting of two domains, the left one with positive magnetization and the right one with negative magnetization (or vice versa). The boundary between them, located between the spins at sites $i$ and $i+1$ in Fig. 10.7, is known as a domain wall, and the energy of the state is

$$E_1 = E_0 + 2J, \tag{10.40}$$

with magnetization

$$M = i - (N + 1 - i) = 2i - N - 1. \tag{10.41}$$

**Fig. 10.6**    Ground state of the one-dimensional Ising model.

$i \quad i+1$

**Fig. 10.7**    Domain wall in the one-dimensional Ising model located between sites $i$ and $i + 1$.

Note that $i$ can take all values from 1 to $N$, so there are $N$ possible locations for the domain wall, and $N$ distinct microstates with energy $E_1$. Hence, the entropy associated with adding a domain wall to the ground state of the one-dimensional Ising model is

$$S_1 = k_B \ln N. \tag{10.42}$$

The different locations of the domain wall also mean that $M$ takes values from $1 - N$ to $N - 1$ in increments of 2. Thus, when we average over all microstates with energy $E_1$, the average magnetization per spin is $m = M/N = 0$.

Now, the entropy of the ground state is $S_0 = 0$, so we can calculate the change in free energy by introducing a single domain wall to the chain:

$$\begin{aligned} \Delta F &= F_1 - F_0 \\ &= E_1 - TS_1 - E_0 + TS_0 \\ &= 2J - k_BT \ln N. \end{aligned} \tag{10.43}$$

We can see that for sufficiently large $N$, $\Delta F < 0$ for any non-zero temperature. This demonstrates that it is always energetically favourable to excite a domain wall in one dimension as we take the thermodynamic limit, and the corresponding average magnetization per spin is $m = 0$, so we do not expect any ordering at non-zero temperatures in one dimension.[10] We will reach the same conclusion when we solve the one-dimensional Ising model exactly.

## 10.3.2 Two Dimensions

We can also consider the effects of domain walls in the two-dimensional Ising model. Consider the model on an $L \times L$ square with two domains and a straight domain wall, as illustrated in Fig. 10.8. There are $L$ spins involved in the domain wall and $L$ possible locations for the ends of the wall, and so a similar argument to the one we presented for one dimension gives the change in free energy for introducing a domain wall as

$$\Delta F = 2JL - k_BT \ln L. \tag{10.44}$$

At large $L$, $L$ grows faster than $\ln L$, so this implies that $\Delta F > 0$ for any non-zero temperature in the thermodynamic limit.

However, this approach underestimates the entropy associated with allowing a domain wall in two dimensions, as it only counts one possible class of domain wall. We can get a better estimate of the entropy associated with introducing a domain wall as follows. First note that there are of order $L$ possible starting points for the domain wall. If we imagine building the domain wall one pair of spins at a time, there are roughly three choices for the domain wall at each step (we ignore pairs of spins near boundaries and the possibility that the domain wall intersects itself), as illustrated in Fig. 10.9.

---

[10] We have shown this result for nearest-neighbour interactions, but it also holds for finite-ranged interactions involving non-neighbouring spins.

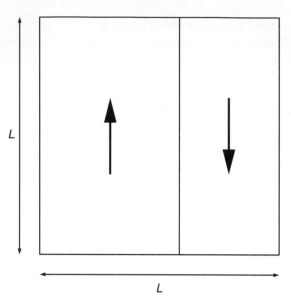

Fig. 10.8   Linear domain wall in the two-dimensional Ising model.

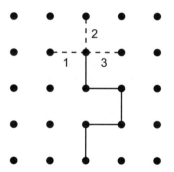

Fig. 10.9   More general domain wall in the two-dimensional Ising model. At each step there are three possibilities for the domain wall (ignoring the possibility that the domain wall intersects itself).

The length of the domain wall will be of order $L$, so there will be roughly $\sim L3^{L}$ domain walls we can introduce to the system; the corresponding entropy will be

$$S \sim Lk_B \ln 3 + k_B \ln L, \tag{10.45}$$

and the free energy change for introducing a domain wall will be

$$\Delta F \sim 2JL - Lk_B T \ln 3 - k_B T \ln L. \tag{10.46}$$

In the limit that $L \to \infty$, this expression implies that $\Delta F < 0$ when

$$k_B T \gtrsim \frac{2J}{\ln 3}. \tag{10.47}$$

This suggests that the ferromagnetic state should be stable up to some non-zero tempera-ture, at which there will be a phase transition. The argument we have presented here can be made much more rigorous to show that there is a non-zero temperature below which there is ferromagnetism. This was first done by Rudolph Peierls (Peierls, 1936).[11]

## 10.4  Exact Solutions of the Ising Model

We have noted the existence of exact solutions to the Ising model in one and two dimensions. We will present a full solution of the one-dimensional model and give a summary of results for the two-dimensional model.

### 10.4.1  Exact Solution in One Dimension

We consider an Ising model on a line with $N$ sites, which are labelled $i = 1, \ldots, N$. If we restrict ourselves to investigating n.n. interactions, then the energy for a spin configuration $\{s_i\}$ is

$$E(\{s_i\}) = -J \sum_{i=1}^{N} s_i s_{i+1} - H \sum_{i=1}^{N} s_i, \tag{10.48}$$

where $s_{N+1}$ is taken to be equal to $s_1$. This is tantamount to assuming periodic boundary conditions, or alternatively, that we are solving the Ising model on a ring (we might reasonably expect that in the limit $N \to \infty$, the answers we get will not depend on the boundary conditions), and has the effect of ensuring that all sites are equivalent since there are no edges to the system.

We can write the partition function as

$$Z_N = \sum_{s_i = \pm 1} \exp\left[ K \sum_{i=1}^{N} s_i s_{i+1} + h \sum_{i=1}^{N} s_i \right], \tag{10.49}$$

where

$$K = \frac{J}{k_B T}; \qquad h = \frac{H}{k_B T}. \tag{10.50}$$

As a result of the assumption that only n.n. spins interact, we can factor the exponential as

$$\exp\left[ K \sum_{i=1}^{N} s_i s_{i+1} + h \sum_{i=1}^{N} s_i \right] = \prod_{i=1,N} \exp\left[ K s_i s_{i+1} + \frac{h}{2}(s_i + s_{i+1}) \right]$$
$$= \prod_{i=1,N} V(s_i, s_{i+1})$$
$$= V(s_1, s_2) V(s_2, s_3) \ldots V(s_{N-1}, s_N) V(s_N, s_1), \tag{10.51}$$

[11] See the book by K. Huang (Huang, 1987) for a detailed account of domain wall arguments demonstrating ordering in the two-dimensional Ising model, along with a discussion of Onsager's solution to the two-dimensional Ising model.

where

$$V(s, s') = \exp\left[Kss' + \frac{h}{2}(s + s')\right]. \tag{10.52}$$

Our particular choice for $V$ is not unique – we could equally well have chosen the term involving $h$ to be $hs$ rather than $\frac{h}{2}(s + s')$, however, the choice we have made has the appealing property that it is symmetric under interchange of $s$ and $s'$:

$$V(s, s') = V(s', s). \tag{10.53}$$

The factorization of the exponential in Eq. (10.51) is a crucial step, since it allows us to write the partition function as a sum over a product of $V$s:

$$Z_N = \sum_{s_i=\pm 1} V(s_1, s_2)V(s_2, s_3) \ldots V(s_{N-1}, s_N)V(s_N, s_1). \tag{10.54}$$

We can regard the values of $V(s, s')$ as elements of a $2 \times 2$ matrix:

$$V = \begin{pmatrix} V(1, 1) & V(1, -1) \\ V(-1, 1) & V(-1, -1) \end{pmatrix} = \begin{pmatrix} e^{K+h} & e^{-K} \\ e^{-K} & e^{K-h} \end{pmatrix}. \tag{10.55}$$

The matrix $V$ is known as a *transfer matrix* – we can think of being transferred one additional site down the chain each time a matrix multiplication is carried out. The summations over $s_2, s_3, \ldots, s_N$ in the expression for the partition function, Eq. (10.54), can be viewed as corresponding to successive matrix multiplications, and the summation over $s_1$ as taking the trace of the matrix that resulted from the multiplications, so we obtain the simple expression that the partition function of the one-dimensional Ising model is just

$$Z_N = \text{Tr}\left[V^N\right]. \tag{10.56}$$

Now, suppose that $\mathbf{x}_1$ and $\mathbf{x}_2$ are the eigenvectors of $V$, so that

$$V\mathbf{x}_i = \lambda_i \mathbf{x}_i \tag{10.57}$$

for each eigenvector, and $\lambda_i$ is the eigenvalue associated with $\mathbf{x}_i$. If we let $P = (\mathbf{x}_1, \mathbf{x}_2)$ be the matrix with columns equal to the eigenvectors of $V$, then

$$VP = P\begin{pmatrix} \lambda_1 & 0 \\ 0 & \lambda_2 \end{pmatrix}, \tag{10.58}$$

and since $V$ is symmetric, it must be possible to choose $\mathbf{x}_1$ and $\mathbf{x}_2$ so that they are orthogonal and linearly independent. This implies that $P^{-1}$ exists, and so

$$V = P\begin{pmatrix} \lambda_1 & 0 \\ 0 & \lambda_2 \end{pmatrix}P^{-1}, \tag{10.59}$$

and hence (using the cyclic property of the trace)[12]

$$Z_N = \text{Tr}\left[\left\{P\begin{pmatrix} \lambda_1 & 0 \\ 0 & \lambda_2 \end{pmatrix}P^{-1}\right\}^N\right] = \text{Tr}\left[\begin{pmatrix} \lambda_1 & 0 \\ 0 & \lambda_2 \end{pmatrix}^N\right] = \lambda_1^N + \lambda_2^N. \tag{10.60}$$

---

[12] For matrices $A$, $B$, and $C$, the cyclic property of the trace is that $\text{Tr}(ABC) = \text{Tr}(CAB) = \text{Tr}(BCA)$. This is most easily shown by writing out the matrix multiplications explicitly:

$$\text{Tr}(ABC) = \sum_{ijk} A_{ij}B_{jk}C_{ki} = \sum_{kij} C_{ki}A_{ij}B_{jk} = \text{Tr}(CAB).$$

If we let $\lambda_1$ be the larger of the two eigenvalues, then we can see that as $N \to \infty$,

$$\lambda_1^N + \lambda_2^N = \lambda_1^N \left[ 1 + \left( \frac{\lambda_2}{\lambda_1} \right)^N \right] \to \lambda_1^N. \tag{10.61}$$

Thus the free energy per spin as $N \to \infty$ is

$$f = \frac{F}{N} = -\frac{k_B T}{N} \ln Z_N \longrightarrow -k_B T \ln \lambda_1. \tag{10.62}$$

Hence, if we diagonalize the $2 \times 2$ matrix $V$ and find the larger of the two eigenvalues, we have an exact expression for the free energy in the limit that $N \to \infty$.

To find the eigenvalues of $V$ we must solve the characteristic equation

$$\det[V - \lambda I] = 0, \tag{10.63}$$

which implies

$$(e^{K+h} - \lambda)(e^{K-h} - \lambda) - e^{-2K} = 0, \tag{10.64}$$

i.e.

$$\lambda^2 - 2e^K \cosh(h)\lambda + 2\sinh(2K) = 0, \tag{10.65}$$

which has the solutions

$$\lambda_\pm = e^K \cosh(h) \pm \left[ e^{2K} \sinh^2(h) + e^{-2K} \right]^{\frac{1}{2}}, \tag{10.66}$$

where we noted $\cosh^2(h) = 1 + \sinh^2(h)$ and used the representation of sinh in terms of exponentials, $\sinh(2K) = \frac{1}{2}\left( e^{2K} - e^{-2K} \right)$, to simplify the expression. We can identify that the larger eigenvalue $\lambda_1$ is $\lambda_+$, and then we have

$$f = -k_B T \ln \left[ e^K \cosh(h) + \left( e^{2K} \sinh^2(h) + e^{-2K} \right)^{\frac{1}{2}} \right]. \tag{10.67}$$

From this expression for the free energy per spin, we can determine the magnetization per spin as

$$m = -\frac{1}{N} \frac{\partial F}{\partial H} = -\frac{1}{k_B T} \frac{\partial f}{\partial h}. \tag{10.68}$$

Evaluating the derivative of $m$ with respect to $h$, we get

$$m = \frac{\frac{\partial}{\partial h} \left[ e^K \cosh(h) + \left( e^{2K} \sinh^2(h) + e^{-2K} \right)^{\frac{1}{2}} \right]}{\left[ e^K \cosh(h) + \left( e^{2K} \sinh^2(h) + e^{-2K} \right)^{\frac{1}{2}} \right]}$$

$$= \frac{e^K \sinh(h) + \dfrac{\sinh(h)\cosh(h)e^{2K}}{\left( e^{2K} \sinh^2(h) + e^{-2K} \right)^{\frac{1}{2}}}}{\left[ e^K \cosh(h) + \left( e^{2K} \sinh^2(h) + e^{-2K} \right)^{\frac{1}{2}} \right]}$$

$$= \frac{e^K \sinh(h)}{\left(e^{2K} \sinh^2(h) + e^{-2K}\right)^{\frac{1}{2}}} \frac{\left[e^K \cosh(h) + \left(e^{2K} \sinh^2(h) + e^{-2K}\right)^{\frac{1}{2}}\right]}{\left[e^K \cosh(h) + \left(e^{2K} \sinh^2(h) + e^{-2K}\right)^{\frac{1}{2}}\right]}$$

$$= \frac{e^K \sinh(h)}{\left(e^{2K} \sinh^2(h) + e^{-2K}\right)^{\frac{1}{2}}}. \tag{10.69}$$

Recalling that $h = H/k_B T$ and $K = J/k_B T$, we can see that if $T$ is finite and we take $H \to 0$, then $h \to 0$ and $m = 0$. Hence there can be no phase transition from a paramagnet to a ferromagnet at any non-zero temperature. In contrast, we know from our considerations earlier that when $T = 0$, $F = U$ and $U$ is minimized by a ferromagnetic configuration, so there is ferromagnetism, but only at $T = 0$.

If we let $H \ll k_B T$, so that $h \ll 1$, and let $t = e^{-2K} = \exp\{-2J/k_B T\}$, then we can write $m$ as

$$m = \frac{t^{-\frac{1}{2}} h}{(t^{-1} h^2 + t)^{\frac{1}{2}}} = \frac{h}{(t^2 + h^2)^{\frac{1}{2}}} = \frac{1}{\left[1 + \left(\frac{t}{h}\right)^2\right]^{\frac{1}{2}}}, \tag{10.70}$$

so that in the vicinity of the critical point at $T = 0$, the magnetization is purely a function of $t/h$, and we can see that $m \to 1$ if we take $t/h \to 0$. This indicates that the limits $H \to 0$ and $T \to 0$ do not commute (since the form of $t$ means that taking $T \to 0$ is equivalent to taking $t \to 0$). If we take $H \to 0$ first, then $m = 0$, whereas if we take $T \to 0$ first, then $m = 1$. Thus the critical temperature for $H = 0$ is $T_c = 0$.

### 10.4.2  Exact Solution in Two Dimensions

It is possible to obtain an exact solution (i.e. an expression for the free energy $f$) to the two dimensional Ising model in two-dimensions when $H = 0$. This was first obtained by Lars Onsager (Onsager, 1944) using a method involving transfer matrices that generalizes the approach in one dimension to two dimensions (and is considerably more complicated than the approach we used for the one-dimensional case), which allowed him to determine that the critical temperature of the two-dimensional Ising model satisfies

$$\sinh\left(\frac{2J}{k_B T_c}\right) = 1. \tag{10.71}$$

Using the definition of sinh in terms of exponentials and solving for $T_c$ gives

$$T_c = \frac{2J}{\ln(1 + \sqrt{2})k_B} \simeq 2.269 \frac{J}{k_B}. \tag{10.72}$$

This compares to our mean field estimate in Eq. (10.23) of $T_c = 4J/k_B$ in two dimensions. The reason for our overestimate is that we treated every spin as experiencing the same ordering field in every microstate, which essentially forces order on the system at a higher temperature than the ordering takes place. The more neighbours each spin has, the more likely that the average spin experiences the average field, and the more accurate mean field theory will be. It is qualitatively wrong in one dimension, predicting ordering at finite temperature when order exists only at $T = 0$, but in two dimensions it is qualitatively correct in predicting ordering, but quantitatively wrong in overestimating the ordering temperature. The quantitative agreement between the mean field prediction for $T_c$ and the exact result improves with increasing dimensionality.

Onsager also obtained an expression for the magnetization as a function of temperature in 1949, and showed that at temperatures just below $T_c$, $m(T)$ has a similar form to the one we determined in mean field theory in Eq. (10.28):

$$m(T) \sim |T_c - T|^\beta, \tag{10.73}$$

but that $\beta = 1/8$ as compared to our mean field estimate of $\beta = 1/2$.

## 10.5  Monte Carlo Simulation of the Ising Model

Even though we do not have an exact solution of the Ising model in three dimensions, the properties of the model have been very well characterized because the model is particularly amenable to being simulated on a computer. We use the Ising model to give a brief account of the application of Monte Carlo methods to simulate the properties of spin models. First we look at the ideas behind Monte Carlo simulations, in particular importance sampling and detailed balance, and then discuss the importance of equilibration so that results do not depend on the choice of initial conditions. We will discuss one particular approach to Monte Carlo simulation, the Metropolis algorithm.

### 10.5.1  Importance Sampling

In the canonical ensemble, the equilibrium probability that the system is in a particular microstate is proportional to its Boltzmann factor, as we saw in Section 4.2.1. Hence for two states, 1 and 2, with energy difference $\Delta E = E_1 - E_2$, the ratio of the probability that the system is in state 1 to the probability that it is in state 2 is

$$\frac{P_1}{P_2} = e^{-\beta \Delta E}. \tag{10.74}$$

Thus, when we calculate equilibrium averages in our Monte Carlo simulation, we will want microstates to be sampled with a probability that depends on $\beta$ and the energy $E(X)$ of a given microstate $X$.

One way we could go about implementing a Monte Carlo simulation would be to generate random configurations and then average over them, attaching importance to each configuration according to the Boltzmann factor. However, this would mean that we would spend most of our time generating high-energy configurations that will contribute to the average with vanishingly small probability, whereas we are usually interested in studying the properties of the Ising model at low temperatures.

A more efficient way to approach Monte Carlo simulation is not to generate statistically independent configurations, but instead to generate configurations by using a Markov chain, so that each new configuration is generated with a probability distribution that depends on the previous configuration. We will not go into the theory of Markov chains, but we are guaranteed to reach the appropriate invariant distribution (in our case the Boltzmann distribution) for sufficiently long times. In practice, after some reasonable number of Monte Carlo steps, one will have a configuration that is statistically independent from a given starting point. We also require that the ratio of probabilities of microstates is respected at every step of the Monte Carlo simulation – this is a very important property, referred to as detailed balance. Below we illustrate a simple Monte Carlo scheme, the Metropolis algorithm, that implements a Markov chain which respects detailed balance and exhibits importance sampling.

### 10.5.2 Metropolis Algorithm

The way that the Metropolis algorithm works is that a move is made to change the state of the system – in the case of the Ising model, this corresponds to flipping a spin (i.e. $s_i \rightarrow -s_i$). The new microstate will differ from the old microstate by an energy $\Delta E$. The probability $P$ that the move is accepted is

$$P = \begin{cases} e^{-\beta \Delta E}, & \Delta E > 0 \\ 1, & \Delta E \leq 0 \end{cases}, \tag{10.75}$$

which clearly respects detailed balance.

In order to implement the Metropolis algorithm in a simulation of the Ising model, we will use the terminology that one Monte Carlo step corresponds to trying to flip all the spins in the lattice once (or once on average) using the Metropolis algorithm. Suppose we try to flip a particular spin at site $i$. We then calculate the energy of the system (or at least the change in energy) after flipping the spin at site $i$. The move is then accepted with a probability given by Eq. (10.75). If the spin flip is rejected, this still counts as a completed trial and we continue to the next spin.

The Metropolis algorithm is not the only method that achieves the correct probability distribution, other closely related popular methods include Glauber dynamics, the heat bath algorithm (both of which have certain advantages in different situations) and various cluster algorithms in which multiple spins are flipped in an individual trial rather than a single spin.

### 10.5.3 Initial Conditions and Equilibration

Unless we have chosen particularly well, it is very likely that the initial state we choose will not be a configuration that has a very high probability at the temperature at which we are interested in solving the model. Hence there will be a certain number of Monte Carlo steps for which the configuration of spins is not representative of the system in equilibrium. This data should not be used for averaging purposes. It can be difficult to determine whether the system has in fact equilibrated. One way that is often effective is to start with two copies of the system, each with different initial conditions, and to consider the system equilibrated when some property of both copies, e.g. the magnetization, agrees to within some pre-determined tolerance. One pair of initial conditions that often yields good results is to have one copy of the system with random spin values (so the initial magnetization per spin $M = 0$), and the other with all of the spins aligned either up or down, which will have $|M| = 1$. At temperatures other than $T = 0$, for a finite system, $0 < |M| < 1$, and hence the two copies of the system will tend to the correct value of the magnetization, one from above, one from below, as illustrated in Fig. 10.10. Once the system is determined to be equilibrated, only one of the replicas needs to be retained. More details on performing Monte Carlo simulations of the Ising model can be found in Problem 10.10. See also Fig. 10.11.

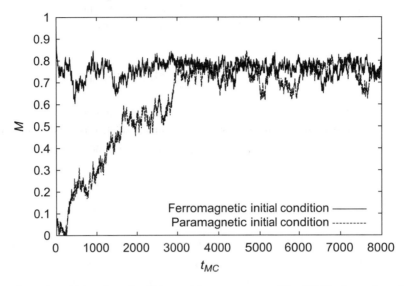

**Fig. 10.10**  Equilibration for a $100 \times 100$ two-dimensional Ising model at a temperature of $T = 2.2\ J/k_B$. Comparison of a ferromagnetic initial condition with a paramagnetic initial condition. Equilibration takes about 3000 Monte Carlo steps.

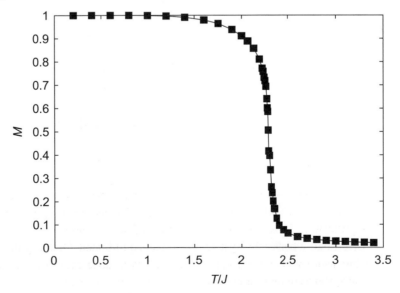

**Fig. 10.11**  Magnetization as a function of temperature in the two-dimensional Ising model for a system of size $100 \times 100$. In a finite-size system, the magnetization at high temperatures does not go exactly to zero, but scales like $\sim 1/\sqrt{N}$.

## 10.6  Connection between the Ising Model and the Liquid–Gas Transition

We have considered the Ising model as a way to understand the transition from a paramagnet to a ferromagnet. The Ising model can also be used to describe aspects of the transition between a liquid and a gas. One possible interpretation of the up and down spins is to consider up spins as atoms and down spins as the absence of atoms. Figures 10.5 and 10.12 illustrate the analogy between the liquid–gas phase diagram and the Ising model phase diagram. The region of magnetization down for the Ising model is mapped to the gas phase ( just a few up spins in a down-spin background is considered to be a low density of atoms, i.e. a gas) and the region of up magnetization (i.e. high density of atoms) is mapped to the liquid phase. In both cases one can go from one phase to the other either by crossing a phase transition line, or by taking a path in the phase diagram that goes around the critical point. There is a second-order phase transition at the critical point, and a line of first-order phase transitions that end at the critical point (i.e. there will be a first-order phase transition on any path that crosses the line).

The connection between spins in a magnetic material and Ising spins is fairly clear, but the analogy between Ising spins and a fluid is less clear. Indeed, the Ising model is not particularly useful for obtaining a quantitative phase diagram for a fluid. However, our observation of power-law behaviour for the magnetization and susceptibility in the Ising model as the temperature $T$ approaches $T_c$ is actually the connection between the two.

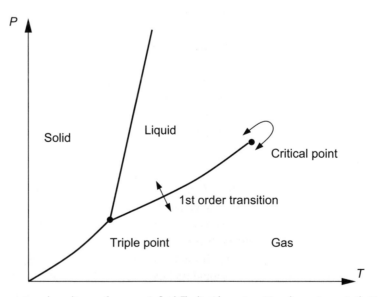

**Fig. 10.12** Pressure–temperature phase diagram for a generic fluid. The liquid–gas transition shares strong similarities with phase transitions in the Ising model.

Near the critical point, analogous quantities in a fluid diverge very similarly to a three-dimensional Ising model, and in fact have the same exponents. This is an example of what is known as *universality*. For systems with the same spatial dimension (i.e. three) and the same symmetry of the order parameter (in both cases the order parameter is Ising-like – up or down for magnetization, large or small for the density), the behaviour near the critical point, i.e. critical exponents, for example $\beta$, $\gamma$ and $\gamma'$ that we found in the Ising model, are conjectured to have the same form, irrespective of the microscopic details of the two phases.

## 10.7 Landau Theory

We conclude with a brief discussion of Landau theory, a very useful framework for understanding ordering and phase transitions from a symmetry point of view. In the Ising model we were able to use mean field theory to find an expression for the free energy as a function of the magnetization per spin $m$ (the order parameter of the ferromagnet). This correctly predicted the existence of a transition in dimensions greater than or equal to two (albeit with quantitatively incorrect transition temperature and critical exponents). Landau built on ideas that we have seen in the example of the Ising ferromagnet to develop a phenomenological theory of second-order phase transitions.

In a second-order phase transition such as the Ising transition, the order parameter grows continuously from zero at the transition as the temperature (or some other intensive variable, e.g. pressure) is lowered. Landau suggested that since the order parameter is zero at the transition and small in its vicinity, the free energy can be expanded as a Taylor series in the order parameter. If we focus on the example of the Ising ferromagnet for clarity, then we know that if we change all up spins to down spins, the energy is unchanged, but the magnetization $m \to -m$. Hence we would expect that only even powers of $m$ are present in the Taylor expansion of the free energy per particle, $f$. If we assume that the free energy is spatially uniform (which is not the case in general, but we ignore it since it complicates our discussion here), then we can write the free energy per spin as

$$f(m, T) = f_0(T) + \alpha(T)m^2 + c(T)m^4 + \cdots . \tag{10.76}$$

Higher-order terms in $m$ (e.g. sixth- order terms) are usually not required to give a good account of important behaviour in the vicinity of $T_c$ for a second-order phase transition. More generally, the idea is to include all terms in the free energy that are compatible with the symmetry of the problem.

We know that at equilibrium the Helmholtz free energy is minimized, so to find the magnetization $m$, we should find the value $\widetilde{m}$ that minimizes $f(m, T)$, hence

$$\left. \frac{\partial f(m, T)}{\partial m} \right|_{m=\widetilde{m}} = 2\alpha\widetilde{m} + 4c\widetilde{m}^3 = 0, \tag{10.77}$$

so if $c > 0$ (if $c$ is negative, then we cannot get a minimum of the free energy at $m = 0$ unless the $m^6$ term has a positive coefficient) then if $\alpha < 0$,

$$\widetilde{m} = \pm\sqrt{-\frac{\alpha}{2c}}, \tag{10.78}$$

whereas $\widetilde{m} = 0$ if $\alpha > 0$. From our knowledge that $\widetilde{m} = 0$ for $T > T_c$ and $\widetilde{m} \neq 0$ for $T < T_c$, we expect that $\alpha > 0$ for $T > T_c$ and $\alpha < 0$ for $T < T_c$, with $\alpha = 0$ at $T = T_c$. Since we are performing an expansion in the vicinity of $T_c$, let

$$\alpha(T) \simeq a_0(T - T_c) + \cdots, \tag{10.79}$$

$$c(T) \simeq c_0 + \cdots, \tag{10.80}$$

in which case we can simplify the free energy expansion to

$$f \simeq f_0(T) + a_0(T - T_c)m^2 + c_0 m^4, \tag{10.81}$$

and our solution above implies that

$$\widetilde{m} \simeq \begin{cases} 0, & T > T_c \\ \left(\dfrac{a_0}{2c_0}\right)^{\frac{1}{2}} (T_c - T)^{\frac{1}{2}}, & T < T_c \end{cases}, \tag{10.82}$$

which we note is exactly the mean field behaviour of the magnetization that we found for the Ising model in Eq. (10.28).

In Fig. 10.13, the mean field free energy of the Ising model, Eq. (10.18), is plotted as a function of $m$ for temperatures above, equal to and below $T_c$, which shows schematically the behaviour captured in Eq. (10.81). At temperatures $T > T_c$ there is a single minimum in the free energy at $\widetilde{m} = 0$, whereas for temperatures below $T_c$, two minima, at $\pm\widetilde{m}$, are present.

### 10.7.1 Symmetry-Breaking Fields

We can also include a magnetic field $H$ in the free energy for the ferromagnet, in which case $f$ is modified to take the form

$$f \simeq f_0(T) + a(T - T_c)m^2 + bm^4 - Hm. \tag{10.83}$$

The magnetic field $H$ couples to the order parameter $m$ and is known as a symmetry-breaking field. Without the magnetic field, the free energy depends only on the magnitude of the magnetization $|m|$, but is insensitive to its direction. With the magnetic field, there is a particular sign of the magnetization which is preferred, and hence the magnetic field *breaks the symmetry* between up and down. When a phase transition such as the Ising transition takes place in the absence of a magnetic field as we discussed earlier, only one of the two minima with equal free energy can be chosen, and this is referred to as *spontaneous symmetry breaking*, since an individual system will choose one particular minimum, but neither minimum is more probable than the other. The long sought-after Higgs boson (discovered in 2015) is believed to arise from spontaneous breaking of electroweak symmetry, and gives rise to the masses of the $W^\pm$ and $Z$ bosons.

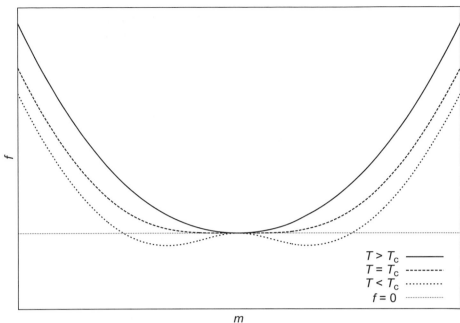

Fig. 10.13 Mean field free energy per spin for the Ising model, Eq. (10.18), as a function of magnetization $m$ at temperatures $T > T_c, T = T_c$ and $T < T_c$. A temperature-dependent piece that is independent of $m$, $f_0(T) = -k_B T \ln 2$, has been subtracted from each of the curves.

If we minimize $f$ with respect to $m$ as before, then we get

$$2a_0(T - T_c)m + 4c_0 m^3 - H = 0, \tag{10.84}$$

and with a further derivative with respect to $H$ before sending $H \to 0$, we get

$$2a_0(T - T_c)\chi + 12c_0 m^2 \chi - 1 = 0. \tag{10.85}$$

We can calculate the magnetic susceptibility for $T > T_c$ very straightforwardly (the $m^2$ term in Eq. (10.85) will give no contribution for $T > T_c$):

$$\chi_M = \frac{1}{2a_0}|T - T_c|^{-1}. \tag{10.86}$$

For $T < T_c$ we can use the expression $m^2 = -\alpha(T)/2c_0$ in Eq. (10.85) to obtain

$$\chi_M = \frac{1}{4a_0}|T - T_c|^{-1}. \tag{10.87}$$

We can also determine from Eq. (10.84) that at $T = T_c$,

$$m \sim \left(\frac{1}{4c_0}\right)^{\frac{1}{3}} H^{\frac{1}{3}}. \tag{10.88}$$

Both the temperature dependence and magnetic field dependence we have determined here are identical to those found by mean field theory, but required considerably less effort to calculate than, for example, our calculation of the susceptibility near $T_c$ for the Ising model.

## 10.7.2  Landau Theory and First-Order Phase Transitions

The Landau theory phenomenology we have explored above is useful in a much wider variety of contexts than Ising ferromagnetism. The general recipe is to construct all terms in the free energy involving the order parameter that are consistent with the symmetry of the problem (this was why we only had even powers of $m$ for a ferromagnet), and this will provide the correct Landau theory. The susceptibility (or its equivalent) can be determined by using a symmetry-breaking field that couples to the order parameter.

Landau theory can also be used to describe first-order phase transitions. If one includes a cubic term in the free energy, e.g.

$$f = f_0 + a(T - T_0)\rho^2 + c\rho^3 + d\rho^4, \tag{10.89}$$

where $\rho$ might correspond to the strength of ordering at a particular wavevector (i.e. this could describe the liquid-to-solid phase transition), then at high temperatures there is still only one minimum of $f$. At lower temperatures (but still higher than $T_c$) there is a local minimum at a different value of the order parameter to the minimum that exists at high temperatures. Eventually at $T_c$ (in general $T_c \neq T_0$) the second minimum becomes lower than the first minimum. This implies that the order parameter will jump discontinuously at the phase transition, rather than growing smoothly as it does in a second-order phase transition. Another free energy that can also lead to a first-order transition is of the form

$$f = f_0 + a(T - T_0)\rho^2 + c\rho^4 + d\rho^6, \tag{10.90}$$

when $c = c_0(T - T^*)$, with $T^* > T_c$, so that $c$ changes sign at some temperature above $T_c$. In these cases $\rho$ may not be small near the transition, and so there is no guarantee that the expansion of $f$ in powers of $\rho$ is a reasonable approximation in the vicinity of the transition, violating the assumptions made in developing Landau theory.

# 10.8  Summary

We introduced the idea that different phases of matter correspond to different broken symmetries. For instance, in a ferromagnet, rotational symmetry is broken by the development of a magnetization, which picks out a particular direction in space. In the Ising model, the up–down symmetry is broken at low temperatures when the magnetization becomes non-zero. Quantities such as the magnetization act as an order parameter and quantify the amount of ordering in a phase where symmetry has been broken. We considered these ideas in the context of the Ising model, and introduced mean field theory as a means to investigate the qualitative behaviour of the model and estimate the critical temperature for ferromagnetic ordering. The ideas developed in the Ising model have wider application, and can be formalized in Landau theory, which provides a framework for investigating ordering for either first or second-order phase transitions.

# Problems

**10.1** It is possible to map a simple model of a polymer onto a spin model. If the starting point of monomer $i$ is at position $\mathbf{X}_i$, then the orientation of the $i^{\text{th}}$ monomer is given by

$$\mathbf{S}_i = \mathbf{X}_{i+1} - \mathbf{X}_i,$$

where we assume that there are $N + 1$ monomers and each monomer has length $a$. To study the elastic properties of the polymer, apply a force $\mathbf{F}$ at $\mathbf{X}_{N+1}$ and a force $-\mathbf{F}$ at $\mathbf{X}_1$. Hence the energy is

$$\mathcal{E} = -(\mathbf{X}_{N+1} - \mathbf{X}_1) \cdot \mathbf{F}.$$

(a) Rewrite the energy in the form of non-interacting spins $\mathbf{S}_i$ in the presence of an effective magnetic field.

(b) Using part (a) or otherwise, show that the length of the polymer takes the form (for a force of magnitude $F$ which is applied along the $z$ direction)

$$L = Na \left( \coth \frac{Fa}{k_B T} - \frac{k_B T}{Fa} \right).$$

Show that in the limit of small force ($Fa \ll k_B T$), this reduces to Hooke's law and find the corresponding spring constant.

**10.2** Consider spin-3/2 atoms for which $S_i^z$ can take the values $-3/2, -1/2, 1/2$ or $3/2$, placed in a magnetic field $B$ and experiencing a crystal field $K$. The energy of the spin system is given by

$$E = B \sum_{i=1}^{N} S_i + K \sum_{i=1}^{N} S_i^2.$$

Calculate the magnetization per spin

$$m = \frac{M}{N} = -\frac{\partial F}{\partial B},$$

and show that in the limit $K \to \infty$, $m$ behaves identically to that found for spin-1/2 spins for which $S_i = \pm 1/2$.

**10.3** The Ising model can be used to describe binary alloys such as $\beta$-brass, which is composed of approximately equal numbers of Cu and Zn atoms. As a simplified model of a binary alloy, consider a square lattice of atoms which can be either of type 1 or type 2. Set spin values $+1$ for type 1 atoms and $-1$ for type 2 atoms and let there be $N_1$ type 1 atoms and $N_2$ type 2 atoms, such that $N_1 + N_2 = N$.

Let the interaction energies between two neighbouring atoms be $E_{11}, E_{22}$ and $E_{12}$, and there be $N_{11}, N_{22}$ and $N_{12}$ bonds of each type. The energy of the binary alloy is thus

$$E_{\text{binary alloy}} = -E_{11} N_{11} - E_{22} N_{22} - E_{12} N_{12}.$$

Show that we can write this in the form

$$E_{\text{Ising}} = -J \sum_{\langle ij \rangle} s_i s_j - B \sum_i s_i - CN,$$

where $CN$ is just a constant shift from the energy of the regular Ising model, and find expressions for $J$, $B$ and $C$.

**10.4** Calculate the energy $U$ and the heat capacity $C_V$ for the Ising model using mean field theory. Find the behaviour of $C_V$ for temperatures just above and just below $T_c$, when the magnetic field $H = 0$.

**10.5** Using the mean field free energy for the Ising model as a starting point, calculate $\alpha(T)$ and $c(T)$ in the Landau expansion of the free energy.

**10.6** Using the mean field free energy for the Ising model, describe the behaviour of $m$ when $H \neq 0$. Show explicitly that when $T = T_c$ and $H$ is close to zero, $m(H) \sim H^\delta$ and find $\delta$.

**10.7** Comment briefly on whether the ergodic hypothesis holds in ferromagnets at temperatures below their Curie temperature.

**10.8** We have used mean field theory to determine the Curie temperature for an Ising ferromagnet. Now consider a model of an Ising antiferromagnet, for which the energy in the absence of a magnetic field is given by

$$E = J \sum_{i \neq j} s_i s_j,$$

with $J > 0$, and where we restrict the sum over $i$ and $j$ to sites $i$ and $j$ that are neighbours.

(a) For antiferromagnets, the symmetry of the lattice matters much more than for ferromagnets. To illustrate this, consider a triangle consisting of three sites so that the energy is given by

$$E = J(s_1 s_2 + s_2 s_3 + s_3 s_1).$$

See Fig. 10.14. Find the lowest-energy arrangement of the spins and the associated energy $E_0$ when:

   (i) $J < 0$, i.e. the interactions are ferromagnetic.
   (ii) $J > 0$, i.e. the interactions are antiferromagnetic.
       Comment on the difference between cases (i) and (ii).
       At non-zero temperature $T$ and for $J < 0$:
   (iii) find the temperature at which the probability of finding the system in an excited state is 50%.

(b) Now consider an antiferromagnetic Ising model on a cubic lattice. In an antiferromagnet at zero temperature, we can divide the lattice into two sublattices, $A$ and $B$, as illustrated in Fig. 10.15 for the case of a square lattice.

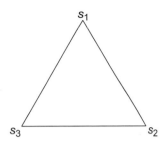

Fig. 10.14 Triangular plaquette with three spins.

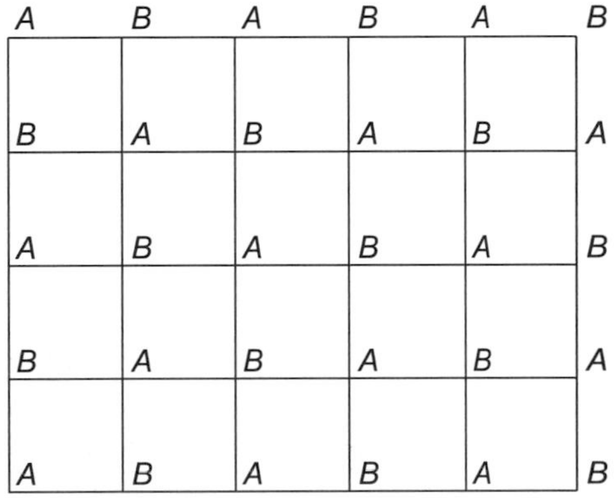

| A | B | A | B | A | B |
|---|---|---|---|---|---|
| B | A | B | A | B | A |
| A | B | A | B | A | B |
| B | A | B | A | B | A |
| A | B | A | B | A | B |

Fig. 10.15    $A$ and $B$ sublattices on a square lattice.

All the neighbours of every spin on sublattice $A$ are on sublattice $B$ and have the opposite spin to the spins on sublattice $A$. This is more complicated than a ferromagnet, but we can still define an order parameter, the staggered magnetization

$$M_s = \left\langle \sum_i (-)^{\eta_i} s_i \right\rangle,$$

where $\eta_i = 1$ if $s_i$ is on sublattice $A$ and $\eta_i = -1$ if $s_i$ is on sublattice $B$.

Using the staggered magnetization per particle $m_s = M_s/N$, and taking the magnetic field to be staggered (so that the magnetic field term in the energy is $\sum_i H(-1)^{\eta_i} s_i$), use mean field theory to determine an estimate for the critical temperature of an Ising antiferromagnet, the Néel temperature. How does this compare to the Curie temperature of an Ising ferromagnet?

**10.9** Consider a model of Heisenberg spins (that you may treat as classical vectors) with n.n. interactions and a field applied in the $z$ direction:

$$H = -J \sum_{\langle ij \rangle} \mathbf{S}_i \cdot \mathbf{S}_j - h \sum_i S_i^z.$$

(a) Use mean field theory to find the critical temperature for this model.

(b) Now add an anisotropy term to the Hamiltonian

$$H_{\text{anisotropy}} = -K \sum_i (S_i^z)^2,$$

and find the equation that the magnetization per spin satisfies (you may assume that $\beta K \ll 1$, where $\beta = 1/k_B T$).

**10.10  Monte Carlo simulation of the two-dimensional Ising model**

Consider a square lattice in two dimensions, with $L$ sites in both the $x$ and $y$ directions. On each site there is a spin $S_i$ which can take values of $\pm 1$. We consider each spin to interact only with its four nearest neighbours (n.n.), so that the total energy is

$$E = -\frac{J}{2} \sum_i \sum_{j \in \{\text{n.n. } i\}} S_i S_j.$$

We will consider $J$ to be positive and set it equal to 1 (this is equivalent to measuring energy in units of $J$).

Write a code to perform the following tasks:

1. Initialize your simulation with two copies of the system, one with the spins randomly assigned, the other with all spins set equal to $+1$. Calculate the initial energies of each copy of the system. Assume periodic boundary conditions, so that for example a spin on the left edge of your system has a spin on the right edge as a neighbour.

2. Perform a sweep through the lattice and use the Metropolis algorithm to update spins (i.e. attempt to flip spin 1, then attempt to flip spin 2, etc. – if there are $N$ spins in the lattice, it is often better to pick a random spin $N$ times (this may lead to some repetition) rather than cycling through in order as this avoids building accidental correlations into your updating scheme). This corresponds to one Monte Carlo time step.

3. Determine when the magnetization of the two copies of the system is close enough to consider the system equilibrated.

4. Analyse a configuration every 50 sweeps after equilibration, and use at least 200 configurations to form an ensemble average. At temperatures close to $T_c$ there is a phenomenon known as critical slowing down, so you may need more than 50 sweeps to get statistically independent configurations.

The thermodynamic properties of the system depend on two quantities: $\beta J$ and $L$. Consider three well-spaced values of $L$, e.g. $L = 10$, 20 and 40.

For each $L$ perform runs at temperatures (in units of $J/k_B$) of 1.5, 2, 2.1, 2.2, 2.25, 2.26, 2.27, 2.28, 2.29, 2.3, 2.4, 2.5 and 3. You will need to find the following for each configuration: the magnetization per spin ($N = L^2$)

$$M = \frac{1}{N} \left| \sum_{i=1}^{N} S_i \right|,$$

(note the absolute value, since we are interested in whether there is ordering or not but do not care about the sign of the magnetization), the energy per spin

$$E = -\frac{1}{N}\frac{J}{2}\sum_{i,j} S_i S_j$$

and also $M^2$, $M^4$ and $E^2$. (Note that it is possible to update the energy without recalculating the full expression for the energy after each spin flip – this will save a lot of time in the simulation.)

From your configurations you can obtain ensemble averages $\langle M \rangle$, $\langle M^2 \rangle$, $\langle E \rangle$ and $\langle E^2 \rangle$. Use these to determine the magnetic susceptibility $\chi$ and the heat capacity $C_V/k_B$. Do you notice anything about the temperature dependence of $\chi$?

# A
## Appendix A  Gaussian Integrals and Stirling's Formula

This appendix gives a derivation of several mathematical results that we will use extensively. The first is formulae for Gaussian integrals, the second is the gamma function and the third is Stirling's formula, which allows us to approximate $N!$ when $N$ is large, the derivation of which relies on knowledge of Gaussian integrals and the gamma function.

## A.1  Gaussian Integrals

A particular class of integrals that occur many times in statistical mechanical calculations (especially involving the Maxwell–Boltzmann velocity distribution) are the Gaussian integrals. The basic Gaussian integral is

$$\int_{-\infty}^{\infty} dx \, e^{-\alpha x^2},$$

which can be evaluated by the following trick of squaring the integral and converting to polar co-ordinates:

$$
\begin{aligned}
\left( \int_{-\infty}^{\infty} dx \, e^{-\alpha x^2} \right)^2 &= \int_{-\infty}^{\infty} dx \, e^{-\alpha x^2} \int_{-\infty}^{\infty} dy \, e^{-\alpha y^2} \\
&= \int_{0}^{\infty} r \, dr \int_{0}^{2\pi} d\theta \, e^{-\alpha r^2} \\
&= 2\pi \int_{0}^{\infty} dr \, r \, e^{-\alpha r^2} \\
&= 2\pi \left[ -\frac{1}{2\alpha} e^{-\alpha r^2} \right]_{0}^{\infty} \\
&= \frac{\pi}{\alpha}.
\end{aligned}
\tag{A.1}
$$

Taking the square root of both sides of the equation we get the desired result:

$$
\int_{-\infty}^{\infty} dx \, e^{-\alpha x^2} = \sqrt{\frac{\pi}{\alpha}}.
\tag{A.2}
$$

For integrals of the form

$$\int_{-\infty}^{\infty} dx \, x^n \, e^{-\alpha x^2},$$

we can note that if $n$ is odd, the integrand is an odd function and the integral will vanish. If $n = 2m$ is an even integer, then if we take a derivative of the original Gaussian integral with respect to $\alpha$, we find the following:

$$-\frac{\partial}{\partial \alpha} \int_{-\infty}^{\infty} dx\, e^{-\alpha x^2} = \int_{-\infty}^{\infty} dx\, x^2\, e^{-\alpha x^2}. \tag{A.3}$$

Hence, from our knowledge of the $m = 0$ integral in Eq. (A.2), we obtain

$$\int_{-\infty}^{\infty} dx\, x^{2m}\, e^{-\alpha x^2} = \left(-\frac{\partial}{\partial \alpha}\right)^m \int_{-\infty}^{\infty} dx\, e^{-\alpha x^2} = \left(-\frac{\partial}{\partial \alpha}\right)^m \sqrt{\frac{\pi}{\alpha}}. \tag{A.4}$$

Finally, if $n$ is an odd integer, and we are interested in an integral from 0 to $\infty$ only, then by letting $y = \alpha x^2$, we can find the following:

$$\int_0^{\infty} dx\, x^{2m+1}\, e^{-\alpha x^2} = \frac{1}{2\alpha^{m+1}} \int_0^{\infty} dy\, y^m\, e^{-y} = \frac{\Gamma(m+1)}{2\alpha^{m+1}} = \frac{m!}{2\alpha^{m+1}}, \tag{A.5}$$

where we used the gamma function, $\Gamma(x)$, which is the generalization of the factorial function to non-integer values, and is discussed in detail in the following section.

## A.2 Gamma Function

The gamma function, defined as

$$\Gamma(\nu) = \int_0^{\infty} dt\, t^{\nu-1} e^{-t}, \tag{A.6}$$

has the property that for $\nu$ an integer $n$,

$$n! = \Gamma(n+1). \tag{A.7}$$

More generally, $\Gamma(\nu + 1) = \nu\Gamma(\nu)$, which we can establish using integration by parts (for $\nu$ non-negative):

$$\begin{aligned}
\Gamma(\nu + 1) &= \int_0^{\infty} dt\, t^{\nu} e^{-t} \\
&= \left[-t^{\nu} e^{-t}\right]_0^{\infty} + \nu \int_0^{\infty} dt\, t^{\nu-1} e^{-t} \\
&= \nu\Gamma(\nu).
\end{aligned} \tag{A.8}$$

Now,

$$\Gamma(1) = \int_0^{\infty} dt\, e^{-t} = 1, \tag{A.9}$$

hence $\Gamma(2) = 1.1 = 1!$, $\Gamma(3) = 2.\Gamma(2) = 2!$, etc.

It is also of interest to note that (using the substitution $y^2 = t$)

$$
\begin{aligned}
\Gamma\left(\frac{3}{2}\right) &= \int_0^\infty dt\, t^{\frac{1}{2}} e^{-t} \\
&= \int_0^\infty dy\, 2y^2 e^{-y^2} \\
&= 2\int_0^\infty dy\, y^2 e^{-y^2} \\
&= \left[-2\frac{\partial}{\partial\alpha} \int_0^\infty dy e^{-\alpha y^2}\right]_{\alpha=1} \\
&= -2\left[\frac{\partial}{\partial\alpha}\frac{1}{2}\sqrt{\frac{\pi}{\alpha}}\right]_{\alpha=1} \\
&= \frac{\sqrt{\pi}}{2},
\end{aligned}
\tag{A.10}
$$

where we used Eq. (A.3) to perform the Gaussian integral.

## A.3  Stirling's Formula

In the limit of large $N$, it is very inconvenient to calculate $N!$ by directly multiplying $N$ integers together. Stirling's formula gives a very accurate way to determine factorials and is most easily derived using the gamma function. Writing

$$
n! = \Gamma(n+1) = \int_0^\infty dt\, t^n e^{-t},
\tag{A.11}
$$

we note that the integrand is sharply peaked when $n$ is large (as illustrated in Fig. A.1 for $n = 40$), and so most of the contribution to the integral comes from the peak region. This motivates the following approach, which is a very simple example of the method of steepest descents. Start by writing

$$
\Gamma(n+1) = \int_0^\infty dt\, t^n e^{-t} = \int_0^\infty dt\, e^{-g(t)},
\tag{A.12}
$$

where

$$
g(t) = t - n\ln(t).
\tag{A.13}
$$

For a general $g(t)$ we can expand around the maximum of $g(t)$ at the value $t = t_0$ as

$$
g(t) = g(t_0) + (t - t_0)g'(t_0) + \frac{1}{2!}(t - t_0)^2 g''(t_0) + \cdots.
\tag{A.14}
$$

Now, we note that in the case of $g(t) = t - n\ln t$:

$$
g'(t) = 1 - \frac{n}{t}; \qquad g''(t) = \frac{n}{t^2},
\tag{A.15}
$$

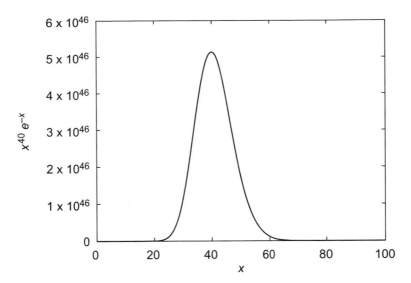

**Fig. A.1**  Integrand of the gamma function for $n = 40$.

hence $t_0 = n$ (as we might have expected from Fig. A.1), and

$$g'(t_0) = 0; \qquad g''(t_0) = \frac{1}{n}, \tag{A.16}$$

so when $n$ is large

$$
\begin{aligned}
\Gamma(n + 1) &\simeq e^{-g(t_0)} \int_0^\infty dt \, e^{-\frac{1}{2}\frac{(t-t_0)^2}{n}} \\
&\simeq e^{-n} e^{n \ln n} \int_{-\infty}^\infty dt \, e^{-\frac{1}{2}\frac{(t-t_0)^2}{n}} \\
&= \sqrt{2\pi n}\, e^{n \ln n - n},
\end{aligned} \tag{A.17}
$$

where in going from the first line to the second line we extended the integral to $-\infty$, which introduces a negligible error since the bulk of the contribution to the integral comes from $t \sim n \gg 0$. This gives us Stirling's approximation to the factorial, also known as **Stirling's formula**:

$$N! \simeq \sqrt{2\pi N}\, N^N e^{-N}, \tag{A.18}$$

which may also be expressed for the logarithm of $N!$ as

$$\ln(N!) \simeq N \ln N - N + \frac{1}{2}\ln(2\pi N). \tag{A.19}$$

This form of the approximation is generally good enough for most work. If we try the expression for $N = 2$ we get the estimate $2! \sim 1.919$, which is very encouraging given that 2 is not particularly large! The approximation works well because of the sharpness of the peak around $t = t_0$. The Gaussian that we integrate over to obtain Eq. (A.17) has a width $\delta t/t \sim 1/\sqrt{n} \to 0$ as $n \to \infty$. To get a more accurate approximation we can keep higher-order terms in the expansion about the maximum in the integrand, in which case we get

$$N! \simeq \sqrt{2\pi N} N^N e^{-N} \left(1 + \frac{1}{12N} + \cdots\right), \qquad (A.20)$$

which for $N = 2$ gives $2! \sim 1.999$.

# B  Appendix B  **Primer on Thermal Physics**

In this book we have generally assumed that the reader is familiar with thermodynamics. However, for the reader who has not seen thermodynamics previously, or for whom a refresher may be helpful, this appendix is intended to give a brief summary of important concepts and results. Thermodynamics predates statistical mechanics and provides a self-consistent framework for describing the relations of thermodynamic variables. In the nomenclature introduced in Chapter 1, we can regard thermodynamics as a theory of macrostates where there is no reference to microscopic details of the state. For a more extensive introduction to the topic of thermodynamics, the reader is encouraged to consult one of the many texts that have been written on the subject.

## B.1  Thermodynamic Equilibrium

A system is said to be in thermodynamic equilibrium when its physical properties are independent of time and it no longer retains any memory of its initial conditions. Usually equilibrium states can be characterized by a small number of macroscopic physical properties, e.g. particle number, volume and temperature. These properties of the equilibrium state are known as state functions.

There are two main classes of questions we shall be interested in for equilibrium states. First, how are different thermodynamic variables related to each other in an equilibrium state? Such relationships are encoded in equations of state. One well-known example is the ideal gas law, which we can write in the form

$$PV - Nk_BT = 0, \tag{B.1}$$

and relates pressure $P$, volume $V$, particle number $N$ and temperature $T$. Second, how do thermodynamic variables change as we go from one equilibrium state to another?

### B.1.1  Reversible and Irreversible Processes

In making a change to a system it is important to distinguish between processes that are reversible and those which are irreversible. The difference between the two can be illustrated with the example of a container of gas that is taken from state 1 which has volume $V_1$ to state 2 which has volume $V_2$, as shown in Fig. B.1. The transition between the two states can be achieved rapidly by compressing the container of gas from $V_1$ to $V_2$, in

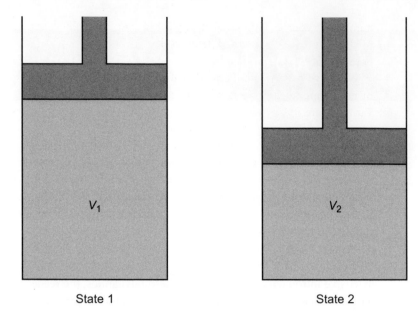

State 1

State 2

**Fig. B.1** Two states of gas in a piston, one with volume $V_1$ and the other with volume $V_2$.

which case the system will go out of equilibrium before establishing a new equilibrium – this is an example of an irreversible change. In a reversible process, the volume is changed very slowly from $V_1$ to $V_2$, so that the gas is always in equilibrium – each small volume change will be accompanied by changes in temperature and pressure so that the system stays in equilibrium. The process is called reversible because if, at some later time, we were to run the process backwards, taking the volume from $V_2$ to $V_1$, we would end up with the gas having the same values for thermodynamic variables as when we started. Reversible processes are also referred to as quasi-static, since they must be carried out slowly enough that equilibrium is maintained at all times. In order that a process be strictly reversible, it should be carried out infinitely slowly, so in practice, reversible processes are an idealization.

Reversible processes may be carried out holding various thermodynamic variables constant. Examples include isobaric processes (constant pressure), isothermal processes (constant temperature), isochoric processes (constant volume) and isentropic processes (constant entropy).

## B.1.2  State Functions

Thermodynamics focuses on macroscopic quantities without reference to microscopic details. We can characterize the state of a system with state functions, which come in two classes: extensive and intensive.

Extensive variables grow with the size of the system – if we take two copies of a system and combine them, then the value of an extensive variable of the combined system will be twice that of the original system. Examples of extensive variables include particle number $N$, volume $V$, internal energy $U$ and entropy $S$. Intensive variables are independent of the size of the system. Examples include pressure $P$, temperature $T$ and chemical potential $\mu$.

When we refer to the thermodynamic state of a system, this means that we specify all of its thermodynamic variables. In an equilibrium state the values of the thermodynamic variables are independent of their initial values (the values depend only on the constraints placed on a system by its environment).

An important feature of thermodynamic variables is that they are functions of state – they are completely determined once the components of the system and the external conditions applied to it are specified. Not all quantities share this property. For instance, although internal energy is a function of state, neither work, nor heat, both of which can change the energy, are – they both depend on the specific path that a system takes between two thermodynamic states.

## Exact Differentials

The property of state functions that they only depend on the state of the system and not the history of how that state was reached (i.e. it is irrelevant whether the equilibrium state was reached via a reversible or an irreversible process) has mathematical consequences. Specifically, state functions have exact differentials. For a function $F$ of two variables, $x$ and $y$, an exact differential has the form

$$dF = f(x, y)dx + g(x, y)dy, \tag{B.2}$$

with the additional conditions

$$f(x, y) = \left.\frac{\partial F}{\partial x}\right|_y, \qquad g(x, y) = \left.\frac{\partial F}{\partial y}\right|_x. \tag{B.3}$$

We also have that the mixed second derivative is independent of the order in which the derivatives are taken:

$$\frac{\partial f(x, y)}{\partial y} = \frac{\partial^2 F(x, y)}{\partial y \partial x} = \frac{\partial^2 F(x, y)}{\partial x \partial y} = \frac{\partial g(x, y)}{\partial x}. \tag{B.4}$$

Equation (B.4) implies that the integral of $dF$ along a path in $x$, $y$ space is independent of the path and equivalently that the integral around a closed loop vanishes:

$$\oint dF = 0. \tag{B.5}$$

## Inexact Differentials

Not all differentials that arise in thermodynamics are exact, i.e. $f$ and $g$ are not related as in Eq. (B.4), and such differentials are known as inexact. Quantities that have inexact differentials include work $đW$ and heat $đQ$, where the bar on the differential indicates that they are inexact. For an inexact differential, the integral along a path depends on the path, so they do not correspond to functions of state.

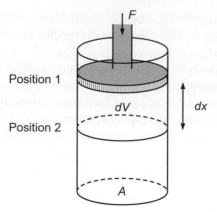

Compression of gas by applying a force $F$ using a piston in a cylinder of cross-sectional area $A$ to achieve a linear displacement $dx$ and volume change $dV$.

## B.1.3   Work and Heat

Work is familiar from mechanics as the energy required to move an object, so for an infinitesimal displacement $d\mathbf{x}$ in the direction of a force $\mathbf{F}$, the work is

$$d W = \mathbf{F} \cdot d\mathbf{x}. \qquad (B.6)$$

We can recast Eq. (B.6) in terms of thermodynamic variables by considering the example illustrated in Fig. B.2. In order to move the piston from position 1 to position 2, a force with magnitude $F$ must be applied to the piston and hence when the piston moves $dx$ downwards, there is an increase in the energy of the gas in the container. If the cross-sectional area of the piston is $A$, then we can write $F = PA$, where $P$ is the pressure of the gas (recalling that pressure is force per unit area). We can also note that the change in the volume of the gas $dV = Adx$, hence we can write the magnitude of the work done on the system by moving the piston in infinitesimal form as $d W = Fdx = PAdx = PdV$.

In this calculation we assumed that the piston is frictionless, so the expression for $d W$ is actually the magnitude of the reversible work. If there is friction it will require work to overcome and will be irreversible work. The final point we need to note is that the volume of the gas decreases as work is done to compress it, hence the inexact differential for reversible work is

$$d W = -PdV. \qquad (B.7)$$

The choice of the negative sign in Eq. (B.7) ensures that the change in energy is positive when work is done on the system and negative when work is done by the system.

We are used to the idea of heat in our everyday lives, for example, sitting near a campfire we feel the transfer of energy from the fire to our skin. This gives the notion that heat is energy flowing from warmer bodies to colder ones. The idea that heat is associated with a flow of energy is important, because it implies that systems do not accumulate heat – there is an energy change associated with adding heat to a system, but we cannot talk about the amount of heat in a system, which makes it clear that heat cannot be a state function. When heat is added to a system, it changes the energy of that system and it is the energy $U$ that is the state function.

We use heat to give a thermodynamic definition of **entropy** in terms of the infinitesimal heat flow at a temperature $T$ so that the infinitesimal change in entropy, $dS$, is defined as

$$dS = \frac{đQ}{T}, \tag{B.8}$$

where $đQ$ refers only to reversible flows of heat and does not include irreversible flows of heat, such as those generated by friction. Focusing on reversible heat flows, the inexact differential for heat is

$$đQ = T\,dS. \tag{B.9}$$

## B.2  The Laws of Thermodynamics

Thermodynamics is a self-consistent framework which follows from the three laws (supplemented with the zeroth law). There are numerous ways to formulate these laws that are equivalent to the statements we give here.

### The Zeroth Law

*If two independent systems are both in thermal equilibrium with a third, then they must also be in thermal equilibrium with each other.*

This law leads to the identification of temperature, $T$, as a quantity that is the same in systems in thermal equilibrium with each other.

### The First Law

*Energy is conserved.*

The energy $U$ of a system is a state function as it measures the total energy of all the degrees of freedom in that system. Changes in the energy can take place either through work being done on or by the system, or via heat flowing into or out of the system. We can write the differential change in energy as

$$dU = đQ + đW. \tag{B.10}$$

The considerations above are sufficient if the number of particles, $N$, in the system is constant. However, even if no heat is added to the system or no work is done on the system, another possible way for the energy to change is for the number of particles in the system to change. One might wish to ascribe this to a form of heat, but this is generally not done, and we can use the differential energy change associated with particles either entering or leaving the system to define the chemical potential $\mu$ through the differential $dU = \mu dN$, where $dN$ is the change in particle number. Thus, using the expressions in Eqs (B.7) and (B.9), and taking into account the possibility that particles enter or leave the system, we can rewrite the differential for the energy as

$$dU = TdS - PdV + \mu dN. \tag{B.11}$$

The derivation of Eq. (B.11) relied on expressions for $dW$ and $dQ$ that were determined for reversible processes. However, because all of the quantities in Eq. (B.11) are state functions, the result must hold independently of the path between two equilibrium states. Hence, Eq. (B.11) holds for irreversible changes as well as reversible ones.

## The Second Law

*The change in global entropy is always non-negative.*

We have chosen to express the second law in terms of entropy, but other formulations may emphasize heat instead, such as:

*Heat does not flow from a cooler to a hotter body.*

However, as we see in Eq. (B.9), there is a close relationship between heat and entropy, so these are equivalent statements. Expressed mathematically, the change in entropy, $\Delta S$, in any process satisfies

$$\Delta S \geq 0, \tag{B.12}$$

where the equality $\Delta S = 0$ only holds for reversible processes. The second law does not preclude the possibility that the entropy of a system connected to an environment decreases, but an entropy decrease in a system must be balanced by a corresponding entropy increase in the environment. It is often helpful to write the second law in the form

$$\Delta S_{\text{total}} = \Delta S_{\text{system}} + \Delta S_{\text{environment}} \geq 0, \tag{B.13}$$

which emphasizes that the total entropy change is positive, but individual components can be negative. For a thermally isolated system, where heat flow into and out of the system is not possible, we must have that

$$\Delta S_{\text{system}} \geq 0. \tag{B.14}$$

## The Third Law

*The entropy per particle goes to zero at zero temperature.*

The third law was originally postulated by Walther Nernst based on the observation that there is no change in entropy in chemical reactions at low temperatures. It can also be understood from a microscopic basis, as discussed in Chapter 4. An important consequence of the third law is that heat capacities go to zero as the temperature goes to zero.

The heat capacity of a system is (in the limit of infinitesimal temperature change $\Delta T$) the heat $\Delta Q$ required to achieve a temperature change $\Delta T$, divided by $\Delta T$:

$$C = \lim_{\Delta T \to 0} \frac{\Delta Q}{\Delta T}. \tag{B.15}$$

Writing $\Delta Q = T\Delta S$ and taking the limit $T \to 0$ gives

$$C = T\frac{\partial S}{\partial T} = \frac{\partial S}{\partial \ln T} \longrightarrow 0, \tag{B.16}$$

which follows from $S \to 0$ and $\ln T \to -\infty$ as $T \to 0$.

# B.3 Thermodynamic Potentials

In Eq. (B.11) we already encountered an example of a thermodynamic potential, the energy $U$, and Eq. (B.11) indicates that $U$ is a function of three independent variables, $S$, $V$ and $N$, so we may write $U(S,V,N)$ to indicate the natural variables for $U$. We can also compare the general expression for the differential of $U$ in terms of the variables $S$, $V$ and $N$ to Eq. (B.11):

$$dU = \frac{\partial U}{\partial S}\bigg|_{V,N} dS + \frac{\partial U}{\partial V}\bigg|_{S,N} dV + \frac{\partial U}{\partial N}\bigg|_{S,V} dN, \tag{B.17}$$

to read off that

$$T = \frac{\partial U}{\partial S}\bigg|_{V,N}, \qquad P = -\frac{\partial U}{\partial V}\bigg|_{S,N}, \qquad \mu = \frac{\partial U}{\partial N}\bigg|_{S,V}. \tag{B.18}$$

Now, the energy $U$ is a function of extensive variables, so it must be true that if we increase the size of the system by a factor of $\lambda$, then

$$U(\lambda S, \lambda V, \lambda N) = \lambda U(S,V,N). \tag{B.19}$$

Hence we may also take a derivative with respect to $\lambda$ on both sides of Eq. (B.19) to obtain

$$U(S,V,N) = \frac{\partial U(\lambda S, \lambda V, \lambda N)}{\partial(\lambda S)}\frac{\partial(\lambda S)}{\partial \lambda} + \frac{\partial U(\lambda S, \lambda V, \lambda N)}{\partial(\lambda V)}\frac{\partial(\lambda V)}{\partial \lambda}$$

$$+ \frac{\partial U(\lambda S, \lambda V, \lambda N)}{\partial(\lambda N)}\frac{\partial(\lambda N)}{\partial \lambda}$$

$$= \frac{\partial U(S,V,N)}{\partial S}S + \frac{\partial U(S,V,N)}{\partial V}V + \frac{\partial U(S,V,N)}{\partial N}N, \tag{B.20}$$

and using the results in Eq. (B.18), we can obtain the fundamental relation

$$U = TS - PV + \mu N. \tag{B.21}$$

In addition to Eq. (B.11), $dU = TdS - PdV + \mu dN$, we can obtain another expression for the differential $dU$:

$$dU = TdS + SdT - PdV - VdP + \mu dN + Nd\mu. \tag{B.22}$$

Equations (B.22) and (B.11) can only both be true if the Gibbs–Duhem relation[1]

$$d\mu = -sdT + vdP \tag{B.23}$$

holds, where $s = S/N$ and $v = V/N$ are the entropy per particle and volume per particle, respectively.

The expression for $U$ in Eq. (B.21) allows us to determine a variety of other relationships between thermodynamic variables. For instance, we can rearrange Eq. (B.21) to get an expression for the entropy $S(U, V, N)$:

$$S = \frac{U}{T} + \frac{PV}{T} - \frac{\mu N}{T}, \tag{B.24}$$

and by rearranging Eq. (B.11) we also have

$$dS = \frac{1}{T}dU + \frac{P}{T}dV - \frac{\mu}{T}dN. \tag{B.25}$$

We can hence read off that

$$\frac{1}{T} = \left.\frac{\partial S}{\partial U}\right|_{V,N}, \tag{B.26}$$

a result that we make extensive use of in Chapter 2, and that

$$P = T \left.\frac{\partial S}{\partial V}\right|_{U,N} \tag{B.27}$$

and

$$\mu = -T \left.\frac{\partial S}{\partial N}\right|_{U,V}. \tag{B.28}$$

In principle, the expressions for $U(S, V, N)$ or $S(U, V, N)$ provide all of the thermodynamic information about a system. However, they are not always the most convenient expressions to work with. For an experiment performed at constant pressure and temperature, it may be easier to work with an expression which is a function of the intensive variables $P$ and $T$ rather than the extensive variables $S$ and $V$. This can be achieved by introducing thermodynamic potentials which are related to the fundamental relations for $U$ and $V$ by a Legendre transform.

---

[1] Also known as the Gibbs–Duhem equation.

## B.3.1 Legendre Transforms and Free Energies

The basic idea of a Legendre transform is that if we have a function $y(x)$, then, instead of representing it as a function of $x$, we can represent it as a function of its derivative $p(x) = dy/dx$. The Legendre transform of the function $y(x)$ with derivative $p(x)$ is

$$Y[p] = y - px. \tag{B.29}$$

We can apply this idea to the energy $U(S, V, N)$ to represent it in terms of either $T$, $P$ or $\mu$, the intensive variables that are conjugate to the extensive variables $S$, $V$ and $N$, respectively.

### Helmholtz Free Energy

Starting from Eq. (B.21), i.e. $U = TS - PV + \mu N$, we can perform the Legendre transformation so that $T$ is the independent variable instead of $S$, in which case we get the **Helmholtz free energy** $F(T, N, V)$:

$$F = U - TS. \tag{B.30}$$

The differential for the Helmholtz free energy is

$$dF = -SdT - PdV + \mu dN, \tag{B.31}$$

from which we can read off the following relations:

$$S = -\left.\frac{\partial F}{\partial T}\right|_{V,N}, \qquad P = -\left.\frac{\partial F}{\partial V}\right|_{T,N}, \qquad \mu = \left.\frac{\partial F}{\partial N}\right|_{T,V}. \tag{B.32}$$

The Helmholtz free energy is minimized at equilibrium when temperature, volume and particle number are held fixed.

### Enthalpy

If we change the independent variable from $V$ to $P$, then a Legendre transformation gives us the **enthalpy** $H(S, V, N)$:

$$H = U + PV. \tag{B.33}$$

The differential for the enthalpy can be determined to be

$$dH = TdS + VdP + \mu dN, \tag{B.34}$$

and the enthalpy is minimized at equilibrium when entropy, pressure and particle number are held fixed.

## Gibbs Free Energy

We can obtain the Gibbs free energy by performing Legendre transformations and changing independent variables for either the Helmholtz free energy (replacing $V$ by $P$) or the enthalpy (replacing $S$ by $T$). The net effect is to change the independent variables from entropy to temperature and volume to pressure, and the expression for the **Gibbs free energy** $G(T, P, N)$ is

$$G = F + PV = H - TS = U - TS + PV. \qquad (B.35)$$

The Gibbs free energy is minimized at equilibrium when temperature, pressure and particle number are held fixed.

## The Grand Potential

The last potential that we shall be interested in is the grand potential, which we can obtain by performing a Legendre transformation on the Helmholtz free energy to change the independent variable from particle number to chemical potential. We will also multiply by a factor of $-1$, which does not change the physical content of the potential, so we have the **grand potential** $\Phi(T, V, \mu)$ as

$$\Phi = \mu N - F. \qquad (B.36)$$

The differential for the grand potential is

$$d\Phi = SdT + PdV + Nd\mu, \qquad (B.37)$$

which implies that we can write

$$S = \left.\frac{\partial \Phi}{\partial T}\right|_{V,\mu}, \qquad P = \left.\frac{\partial \Phi}{\partial V}\right|_{T,\mu}, \qquad N = \left.\frac{\partial \Phi}{\partial \mu}\right|_{T,V}. \qquad (B.38)$$

Due to the factor of $-1$ that we introduced above, the grand potential is maximized rather than minimized at equilibrium when temperature, volume and chemical potential are held fixed.

# B.4  Maxwell Relations

We saw in Eq. (B.4) that for an exact differential

$$dF = \left.\frac{\partial F}{\partial x}\right|_{y} dx + \left.\frac{\partial F}{\partial y}\right|_{x} dy, \qquad (B.39)$$

it is true that

$$\frac{\partial^2 F}{\partial x \partial y} = \frac{\partial^2 F}{\partial y \partial x}. \qquad (B.40)$$

We can use this result to obtain additional relationships between thermodynamic variables. For instance, when $N$ is fixed (i.e. $dN = 0$) we have

$$dU = T\,dS - P\,dV,\qquad\text{(B.41)}$$

and applying the condition on second derivatives, Eq. (B.40), we see that

$$\left.\frac{\partial T}{\partial V}\right|_{S,N} = -\left.\frac{\partial P}{\partial S}\right|_{V,N},\qquad\text{(B.42)}$$

which is an example of a **Maxwell relation**. Focusing on fixed $N$, in addition to Eq. (B.42) which comes from the energy (via $dU$), we can also obtain Maxwell relations from the Helmholtz free energy (via $dF$):

$$\left.\frac{\partial S}{\partial V}\right|_{T} = \left.\frac{\partial P}{\partial T}\right|_{V},\qquad\text{(B.43)}$$

the enthalpy (via $dH$):

$$\left.\frac{\partial T}{\partial P}\right|_{S} = \left.\frac{\partial V}{\partial S}\right|_{P},\qquad\text{(B.44)}$$

and the Gibbs free energy (via $dG$):

$$\left.\frac{\partial S}{\partial P}\right|_{T} = -\left.\frac{\partial V}{\partial T}\right|_{P}.\qquad\text{(B.45)}$$

## B.4.1  Useful Partial Derivative Relations

In order to apply Maxwell relations, it is useful to recall several relations involving partial derivatives for functions of the form $x = x(y, z)$. The first of these is the **chain rule**

$$\left.\frac{\partial x}{\partial y}\right|_{z} = \left.\frac{\partial x}{\partial u}\right|_{z}\left.\frac{\partial u}{\partial y}\right|_{z}.\qquad\text{(B.46)}$$

The second comes from using the expression for the differential for $x$ in terms of $y$ and $z$:

$$dx = \left.\frac{\partial x}{\partial y}\right|_{z}dy + \left.\frac{\partial x}{\partial z}\right|_{y}dz,\qquad\text{(B.47)}$$

and the differential for $z$ in terms of $x$ and $y$:

$$dz = \left.\frac{\partial z}{\partial y}\right|_{x}dy + \left.\frac{\partial z}{\partial x}\right|_{y}dx.\qquad\text{(B.48)}$$

Substitute Eq. (B.48) into Eq. (B.47) to get

$$dx = \left\{\left.\frac{\partial x}{\partial y}\right|_{z} + \left.\frac{\partial x}{\partial z}\right|_{y}\left.\frac{\partial z}{\partial y}\right|_{x}\right\}dy + \left.\frac{\partial x}{\partial z}\right|_{y}\left.\frac{\partial z}{\partial x}\right|_{y}dx,\qquad\text{(B.49)}$$

and in order for the left-hand side and the right-hand side of Eq. (B.49) to be equal, the coefficient of the $dy$ term must vanish, so we obtain

$$\left.\frac{\partial x}{\partial y}\right|_z = -\left.\frac{\partial x}{\partial z}\right|_y \left.\frac{\partial z}{\partial y}\right|_x. \tag{B.50}$$

## B.4.2  Example: Relationship between $C_V$ and $C_P$

We saw in Eq. (B.16) that the heat capacity is given by

$$C = T\frac{dS}{dT}. \tag{B.51}$$

Two common forms of the heat capacity measured experimentally are the heat capacity at constant volume $C_V$ and the heat capacity at constant pressure $C_P$. We can use a Maxwell relation and Eq. (B.50) to find a relation between $C_V$ and $C_P$. Start with the differential for the entropy in terms of $T$ and $V$:

$$dS = \left.\frac{\partial S}{\partial T}\right|_V dT + \left.\frac{\partial S}{\partial V}\right|_T dV, \tag{B.52}$$

and hence the derivative of $S$ with respect to $T$ at constant pressure, multiplied by $T$, is

$$T\left.\frac{\partial S}{\partial T}\right|_P = T\left.\frac{\partial S}{\partial T}\right|_V + T\left.\frac{\partial S}{\partial V}\right|_T \left.\frac{\partial V}{\partial T}\right|_P, \tag{B.53}$$

which we can rewrite as

$$C_P = C_V + T\left.\frac{\partial P}{\partial T}\right|_V \left.\frac{\partial V}{\partial T}\right|_P, \tag{B.54}$$

where we used the Maxwell relation, Eq. (B.43), to rewrite the second term on the right-hand side of Eq. (B.54). We can apply Eq. (B.50) to write

$$\left.\frac{\partial P}{\partial T}\right|_V = -\left.\frac{\partial P}{\partial V}\right|_T \left.\frac{\partial V}{\partial T}\right|_P, \tag{B.55}$$

and then we obtain

$$C_P = C_V - T\left.\frac{\partial P}{\partial V}\right|_T \left(\left.\frac{\partial V}{\partial T}\right|_P\right)^2. \tag{B.56}$$

In the case of an ideal gas where $PV = Nk_BT = nRT$, this simplifies to the relation

$$C_P = C_V + Nk_B = C_V + nR. \tag{B.57}$$

In Chapter 9, we asserted that the heat capacity of an ideal Bose gas has a cusp at the temperature $T_0$ at which Bose–Einstein condensation takes place. In this appendix, we show explicitly how this result arises. We will make use of the notation we previously introduced for the fugacity, $z = e^{\beta\mu}$, and note that at temperatures above $T_0$, we can write the particle number as

$$
\begin{aligned}
N &= \int_0^\infty d\epsilon \frac{g(\epsilon)}{e^{\beta(\epsilon-\mu)} - 1} \\
&= \frac{g_s V}{4\pi^2} \left(\frac{2m}{\hbar^2}\right)^{\frac{3}{2}} \int_0^\infty d\epsilon \frac{\sqrt{\epsilon}}{z^{-1}e^{\beta\epsilon} - 1} \\
&= \frac{g_s V}{4\pi^2} \left(\frac{2m}{\hbar^2}\right)^{\frac{3}{2}} \left(\frac{1}{\beta}\right)^{\frac{3}{2}} \int_0^\infty dx \frac{\sqrt{x}}{z^{-1}e^x - 1} \\
&= \frac{g_s V}{4\pi^2} \left(\frac{2mk_B T}{\hbar^2}\right)^{\frac{3}{2}} \int_0^\infty dx \sqrt{x} \sum_{j=0}^\infty z^{j+1} e^{-(j+1)x} \\
&= g_s V \left(\frac{mk_B T_0}{2\pi\hbar^2}\right)^{\frac{3}{2}} \sum_{j=1}^\infty \frac{z^j}{j^{\frac{3}{2}}} \\
&= g_s n_Q V g_{\frac{3}{2}}(z),
\end{aligned}
\tag{C.1}
$$

where we have defined

$$
g_\nu(z) = \sum_{j=1}^\infty \frac{z^j}{j^\nu}.
\tag{C.2}
$$

It is helpful to note a few properties of the $g$ function. First, in the limit that $z \to 1$:

$$
\lim_{z\to 1} g_\nu(z) = \zeta(\nu),
\tag{C.3}
$$

for values of $\nu$ for which the Riemann zeta function is defined (Fig. C.1), which follows from the definition in Eq. (C.2). Seccond, some special cases that are worth mentioning are

$$
g_1(z) = z + \frac{z^2}{2} + \frac{z^3}{3} + \frac{z^4}{4} + \cdots = -\ln(1 - z)
\tag{C.4}
$$

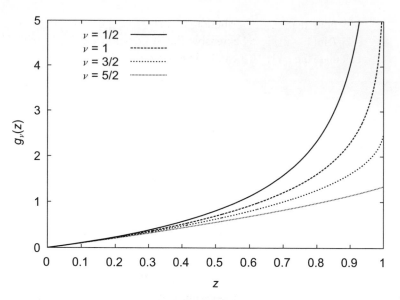

Fig. C.1 Behaviour of the function $g_\nu(z)$ in the range $0 \le z \le 1$ for several values of $\nu$.

for $z < 1$ and

$$g_{\frac{1}{2}}(z) = z + \frac{z^2}{\sqrt{2}} + \frac{z^3}{\sqrt{3}} + \cdots , \qquad (C.5)$$

which diverges as $z \to 1$. We will also need an expression for the entropy, which we can obtain once we have an expression for the grand potential, and we include the contribution from the lowest-energy orbital, which is excluded in the continuum approximation:

$$\Phi = -k_B T \int_0^\infty d\epsilon\, g(\epsilon) \ln\left(1 - ze^{-\beta\epsilon}\right) - k_B T \ln(1 - z)$$

$$= -k_B T \frac{g_s V}{4\pi^2} \left(\frac{2m}{\hbar^2}\right)^{\frac{3}{2}} \int_0^\infty d\epsilon\, \sqrt{\epsilon}\, \ln\left(1 - ze^{-\beta\epsilon}\right) - k_B T \ln(1 - z)$$

$$= -k_B T \frac{g_s V}{4\pi^2} \left(\frac{2m}{\hbar^2}\right)^{\frac{3}{2}} \left\{ \left[\frac{2}{3}\epsilon^{\frac{3}{2}} \ln\left(1 - ze^{-\beta\epsilon}\right)\right]_0^\infty + \frac{2}{3}z\beta \int_0^\infty d\epsilon\, \epsilon^{\frac{3}{2}} \frac{e^{-\beta\epsilon}}{1 - ze^{-\beta\epsilon}} \right\}$$

$$\quad - k_B T \ln(1 - z)$$

$$= -\frac{2}{3} \frac{g_s V}{4\pi^2} \left(\frac{2m}{\hbar^2}\right)^{\frac{3}{2}} \frac{1}{\beta^{\frac{5}{2}}} \int_0^\infty dx\, x^{\frac{3}{2}} \sum_{j=0}^\infty z^{j+1} e^{-(j+1)x} - k_B T \ln(1 - z)$$

$$= -\frac{2}{3} \frac{g_s V}{4\pi^2} \left(\frac{2mk_B T}{\hbar^2}\right)^{\frac{3}{2}} k_B T\, \Gamma\left(\frac{5}{2}\right) g_{\frac{5}{2}}(z) - k_B T \ln(1 - z)$$

$$= -g_s n_Q V k_B T g_{\frac{5}{2}}(z) - k_B T \ln(1 - z). \qquad (C.6)$$

We note that the contribution to the grand potential from the lowest-energy orbital, $-k_B T \ln(1 - z)$, is subextensive (i.e. it grows more slowly with $N$ than $N$), so when we calculate the entropy it can be ignored and we have

$$S = -\frac{\partial \Phi}{\partial T}$$

$$= g_s V k_B n_Q g_{\frac{5}{2}}(z) + \frac{3}{2} g_s V k_B T \frac{n_Q}{T} g_{\frac{5}{2}}(z) + g_s n_Q V k_B T \frac{\partial z}{\partial T} \frac{\partial}{\partial z} g_{\frac{5}{2}}(z)$$

$$= \frac{5}{2} g_s V k_B n_Q g_{\frac{5}{2}}(z) - \frac{g_s V k_B T}{k_B T^2} n_Q \mu z \sum_{j=1}^{\infty} \frac{j z^{j-1}}{j^{\frac{5}{2}}}$$

$$= \frac{5}{2} g_s V k_B n_Q g_{\frac{5}{2}}(z) - \frac{n_Q V \mu}{T} g_{\frac{3}{2}}(z). \tag{C.7}$$

## C.1 Heat Capacity

We know that the heat capacity at constant volume and particle number is given by

$$C_{V,N} = T \left. \frac{\partial S}{\partial T} \right|_{V,N}, \tag{C.8}$$

and we can use the differential for $dS$ that

$$dS = \left. \frac{\partial S}{\partial T} \right|_{V,\mu} dT + \left. \frac{\partial S}{\partial \mu} \right|_{V,T} d\mu, \tag{C.9}$$

to rewrite Eq. (C.8) as

$$C_{V,N} = T \left. \frac{\partial S}{\partial T} \right|_{V,\mu} + T \left. \frac{\partial S}{\partial \mu} \right|_{V,T} \left. \frac{\partial \mu}{\partial T} \right|_N, \tag{C.10}$$

and since $N = N(T, \mu)$ we can also use the differential

$$dN = \left. \frac{\partial N}{\partial T} \right|_{\mu} dT + \left. \frac{\partial N}{\partial \mu} \right|_T d\mu, \tag{C.11}$$

to obtain

$$\left. \frac{\partial \mu}{\partial T} \right|_N = -\frac{\left. \frac{\partial N}{\partial T} \right|_{\mu}}{\left. \frac{\partial N}{\partial \mu} \right|_T}, \tag{C.12}$$

and noting that

$$\frac{\partial^2 \Phi}{\partial T \partial \mu} = -\left. \frac{\partial S}{\partial \mu} \right|_T = -\left. \frac{\partial N}{\partial T} \right|_{\mu}, \tag{C.13}$$

we may write

$$\frac{\partial \mu}{\partial T}\bigg|_N = -\frac{\frac{\partial S}{\partial \mu}\big|_T}{\frac{\partial N}{\partial \mu}\big|_T}. \tag{C.14}$$

Thus

$$C_{V,N} = C_{V,\mu} - T\frac{\left(\frac{\partial S}{\partial \mu}\big|_T\right)^2}{\frac{\partial N}{\partial \mu}\big|_T}, \tag{C.15}$$

and in Eq. (C.7) we showed that the entropy is given by

$$S = \frac{5}{2}Vk_B n_Q g_{\frac{5}{2}}(z) - n_Q V\frac{\mu}{T}g_{\frac{3}{2}}(z), \tag{C.16}$$

hence we have

$$\begin{aligned}C_{V,\mu} &= T\frac{\partial S}{\partial T}\bigg|_{\mu,V} \\ &= \frac{15}{4}k_B n_Q V g_{\frac{5}{2}}(z) - 3n_Q V\frac{\mu}{T}g_{\frac{3}{2}}(z) + n_Q V\frac{\mu^2}{k_B T^2}g_{\frac{1}{2}}(z),\end{aligned} \tag{C.17}$$

$$\frac{\partial S}{\partial \mu}\bigg|_T = \frac{3}{2}\frac{n_Q V}{T}g_{\frac{3}{2}}(z) - n_Q V\frac{\mu}{k_B T^2}g_{\frac{1}{2}}(z) \tag{C.18}$$

and

$$\frac{\partial N}{\partial \mu}\bigg|_T = \frac{n_Q V}{k_B T}g_{\frac{1}{2}}(z). \tag{C.19}$$

Substituting Eqs (C.17), (C.18) and (C.19) in Eq. (C.15) gives

$$C_{V,N} = \frac{15}{4}k_B n_Q V g_{\frac{5}{2}}(z) - \frac{9}{4}k_B n_Q V\frac{g_{\frac{3}{2}}(z)^2}{g_{\frac{1}{2}}(z)}, \tag{C.20}$$

which can be seen to be continuous as $T \to T_0$, i.e. $z \to 1$, even though $g_{\frac{1}{2}}(z)$ diverges as $z \to 1$. However, the second term in Eq. (C.20) leads to a discontinuity in $dC_V/dT$ at $T = T_0$, which is the heat capacity cusp identified in Chapter 9.

# References

Anderson, M. H., Ensher, J. R., Matthews, M. R., Wieman, C. E. and Cornell, E. A. 1995. Observation of Bose–Einstein condensation in a dilute atomic vapor. *Science*, **269**, 198–201.

Bergemann, C., Mackenzie, A. P., Julian, S. R., Forsythe, D. and Ohmichi, E. 2003. Quasi-two-dimensional Fermi liquid properties of the unconventional superconductor $Sr_2RuO_4$. *Adv. Phys.*, **52**, 639–725.

Fixsen, D. J., Cheng, E. S., Gales, J. M., Mather, J. C., Shafer, R. A. and Wright, E. L. 1996. The cosmic microwave background spectrum from the full COBE FIRAS data set. *Astrophys. J.*, **473**, 576–587.

Hill, R. W. and Lounasmaa, O. V. 1957. The specific heat of liquid helium. *Phil. Mag.*, **2**, 143–148.

Huang, K. 1987. *Statistical Mechanics* (2nd edn). Wiley: New York.

Imbrie, J. Z. 2016. On many-body localization for quantum spin chains. *J. Stat. Phys.*, **163**, 998–1048.

Landauer, R. 1961. Irreversibility and heat generation in the computing process. *IBM J. Res. Dev.*, **5**, 183–191.

Langevin, P. 1908. Sur la théorie du mouvement Brownien. *C. R. Acad. Sci. Paris*, **146**, 530–533.

Onsager, L. 1944. Crystal statistics. I. A two-dimensional model with an order–disorder transition. *Phys. Rev.*, **65**, 117–149.

Osheroff, D. D., Richardson, R. C. and Lee, D. M. 1972. Evidence for a new phase of solid He-3. *Phys. Rev. Lett.*, **28**, 885–888.

Peierls, R. 1936. On Ising's model of ferromagnetism. *Math. Proc. Camb. Phil. Soc.*, **32**, 477–481.

Pippard, A. B. 1957. An experimental determination of the Fermi surface in copper. *Phil. Trans. R. Soc. Lond., Ser. A*, **250**, 325–357.

Stedman, R., Almqvist, L. and Nilsson, G. 1967. Phonon-frequency distributions and heat capacities of aluminum and lead. *Phys. Rev.*, **162**, 549–557.

# Index